CAMBRIDGE MONOGRAPHS ON PHYSICS

ARTIFICIAL RADIOACTIVITY

ARTIFICIAL RADIOACTIVITY

BY

P. B. MOON, F.R.S.
Professor of Physics in the University of Birmingham

CAMBRIDGE
AT THE UNIVERSITY PRESS
1949

CAMBRIDGE
UNIVERSITY PRESS

University Printing House, Cambridge CB2 8BS, United Kingdom

Published in the United States of America by Cambridge University Press, New York

Cambridge University Press is part of the University of Cambridge.

It furthers the University's mission by disseminating knowledge in the pursuit of education, learning and research at the highest international levels of excellence.

www.cambridge.org
Information on this title: www.cambridge.org/9781107643932

First published 1949
First paperback edition 2014

A catalogue record for this publication is available from the British Library

ISBN 978-1-107-64393-2 Paperback

GENERAL PREFACE

The Cambridge Physical Tracts, out of which this series of Monographs has developed, were planned and originally published in a period when book production was a fairly rapid process. Unfortunately, that is no longer so, and to meet the new situation a change of title and a slight change of emphasis have been decided on. The major aim of the series will still be the presentation of the results of recent research, but individual volumes will be somewhat more substantial, and more comprehensive in scope, than were the volumes of the older series. This will be true, in many cases, of new editions of the Tracts, as these are re-published in the expanded series, and it will be true in most cases of the Monographs which have been written since the War or are still to be written.

The aim will be that the series as a whole shall remain representative of the entire field of pure physics, but it will occasion no surprise if, during the next few years, the subject of nuclear physics claims a large share of attention. Only in this way can justice be done to the enormous advances in this field of research over the War years.

N. F.

D. S.

May, 1948.

TABLE OF CONTENTS

AUTHOR'S PREFACE

This book is intended to provide an outline of the main phenomena and techniques of radioactivity, as met with in the study of the light and medium-weight radioactive nuclei that can now be made in such great variety.

In writing it, I have borne in mind the reader who wishes to be put into contact with the recent literature of the subject, and have given references mainly to recent papers from which earlier ones may be traced. In these references, the titles of some well-known journals have been abbreviated to initials only; they are listed and explained on the page that follows. The editors and publishers have been good enough to allow me to take account, at the proof stage, of some papers published in 1948.

The radioactive properties of a good many individual nuclei are discussed by way of illustration, but a complete list is impracticable in a book of this size; in any event, the subject is developing so fast that such lists rapidly become incomplete.

I am most grateful to my colleague Professor R. E. Peierls for advice on some theoretical points, and to Professor N. Feather, editor of this series of monographs, for my rescue from many errors.

P. B. M.

BIRMINGHAM,
December, 1948.

LIST OF PERIODICALS

With abbreviations used in this monograph

AMAF,	*Arkiv för Matematik, Astronomi och Fysik.*
AP,	*Annales de Physique.*
CR,	*Comptes Rendus (l'Académie des Sciences, Paris).*
HPA,	*Helvetica Physica Acta.*
KDVS,	*Kgl. Danske Videnskabernes Selskab, Mathematisk-fysiske Meddelelser.*
N,	*Nature.*
P,	*Physica.*
PCPS,	*Proceedings of the Cambridge Philosophical Society.*
PPS,	*Proceedings of the Physical Society.*
PM,	*Philosophical Magazine.*
PR,	*Physical Review.*
PRS,	*Proceedings of the Royal Society (series A).*
RMP,	*Reviews of Modern Physics.*
RSI,	*Review of Scientific Instruments.*
ZP,	*Zeitschrift für Physik.*

CHAPTER I

INTRODUCTION

'Artificial radioactivity' is a convenient abbreviation for 'the radioactivity of certain artificially-made atomic nuclei', and it cannot be emphasized too soon or too strongly that, once such a nucleus has been made, its disintegration is just as spontaneous, and just as immune from outside interference, as that of a naturally occurring radioactive nucleus. A monograph dealing almost exclusively with artificial radioactivity may, however, be justified on several grounds. The natural radioactive substances have mostly been known for nearly half a century and are discussed in many treatises; they belong almost entirely to the upper end of the periodic table of elements, and, owing to the phenomenon of α-activity, they present a characteristic pattern of successive transformations that is absent among the lighter nuclei which are the subject of this book.

Before the discovery of artificial radioactivity, forty-two kinds of radioactive nucleus were known, of which thirty-nine are those of elements having atomic numbers 81, 82, 83, 84, 86, 88, 89, 90, 91 and 92. The large number of radioactive species in this part of the table (39 species among 10 chemical elements) showed that chemically identical atoms, having the same outer electronic structure and therefore the same nuclear electric charge, can have different nuclear structures. Radon, thoron and actinon, for example, are *isotopes* having atomic number (Z) equal to 86; that is to say, their nuclei contain 86 protons each, but they contain 136, 134 and 133 neutrons respectively and their mass numbers (A) are 222, 220 and 219.

Among these thirty-nine heavy radioactive bodies, nineteen disintegrate by emitting an α-particle (helium nucleus), and thereby lose two units of nuclear charge and four of mass; thirteen emit a β-particle (negative electron), gaining one unit of charge without alteration of mass number. The remainder appear to have both roads open to them; this is certain for the bismuth isotopes

RaC, ThC, AcC and RaE † where with ThC, for example, about thirty-five of every hundred identical ThC nuclei emit an α-particle and sixty-five emit a β-particle.

All these nuclei exist in nature because of the extremely long lives of thorium (Th^{232}) and of the uranium isotopes U^{235} and U^{238}, from one or other of which each of the rest descends by successive α- and β-transformations.

With three or four exceptions, such as the very long-lived β-active K^{40}, the naturally occurring isotopes of elements below $Z = 81$ are all stable, but there are often vacancies among (and always, of course, above and below) the existing isotopes of any one element. For example, Cu^{63} and Cu^{65} are found, but not Cu^{64}; N^{14} and N^{15} exist, but not N^{13} or N^{16}. This book is concerned with the nuclei that fill these gaps and particularly with their modes of spontaneous transformation. Because these nuclei are unstable and are not replaced in nature by the disintegration of long-lived ancestors, they do not exist naturally and must be made from stable nuclei by nuclear reactions.

The first artificial nuclear transmutation, made by Rutherford in 1919, resulted from the entry of a fast α-particle into the N^{14} nucleus and the consequent ejection of a proton. The reaction may be written

$$N_7^{14} + He_2^4 = O_8^{17} + H_1^1$$

or, more concisely, $N^{14}(\alpha, p)O^{17}$, where the brackets enclose symbols for the entering and emergent particles. O^{17} is stable, as also are the products of the first transmutation produced with artificially-accelerated protons by Cockcroft and Walton; but by bombarding light elements with α-particles and then presenting the targets to a Geiger-Müller electron counter, Curie and Joliot discovered that in several instances the product is radioactive. A typical example is the reaction $B^{10}(\alpha, n)N^{13}$, following which the N^{13} nuclei disintegrate spontaneously with the emission, as it so happens, of *positive* electrons. The original nuclear reaction is virtually instantaneous while, in contrast, the average lifetime of an N^{13} nucleus is a matter of minutes; but a more fundamental distinction between the nuclear reaction and the radioactive transformation is that the neutron is emitted from a strongly-

† For the evidence concerning RaE, see *PRS*, 190, 20.

excited compound nucleus (N^{14}) formed at the impact, while the positron is emitted from a nucleus that has had time to get rid of spare energy by γ-radiation.

About 500 radioactive nuclear species have now been obtained as the direct or indirect products of reactions in which the projectile and the ejected particle (if any) vary widely in nature and energy. Among these reactions (d, p), (n, γ), (d, n) and (p, n) transformations are the most prominent (here d stands for deuteron, and γ indicates that no material particle, but only γ-radiation, leaves the highly excited compound nucleus—that is to say, that the projectile is simply captured). Books such as Pollard and Davidson's *Applied Nuclear Physics* (New York, Wiley; London, Chapman and Hall) survey these reactions and the techniques by which they have been studied; here, only the radioactive processes will be discussed, and (though exceptions will be made on occasion) radioactive disintegrations of nuclei beyond $Z = 80$ will be left out of account. Alpha-activity and spontaneous fission are accordingly not considered, even though an α-active samarium (probably Sm^{152}_{62}) exists (*PR*, 73, 1125; see also *N*, 158, 197).

The interpretation of the observed processes of artificial radioactivity is based upon the same rules and principles that serve for other domains of nuclear and atomic physics. Without attempting any thorough survey, we may note the following:

1. The conservation, and the quantization in units of $\pm 1{\cdot}60 \times 10^{-19}$ coulomb, of electric charge.

2. The conservation of linear momentum.

3. The conservation of energy, and the equivalence of mass and energy, the total energy W of any system being equal to its mass M multiplied by the square of the velocity of light, c. We may thus write $W = Mc^2 = M_0c^2 +$ kinetic energy, where M_0 is the mass of the system when its centre of gravity is at rest. One atomic mass unit ($\frac{1}{16}$ of the mass of an O^{16} atom) is equivalent to about 931 million electron volts (MeV.) of energy, while the mass of an electron is equivalent to 0·511 MeV.

4. The limitation of nuclear configurations to discrete energy 'levels', as evidenced by the monoenergetic gamma-rays emitted by a nucleus in transitions from one level to another, and the

monoenergetic alpha-rays which involve transitions between levels of a parent and a product nucleus. The difficulty of reconciling this evidence with the continuous spread of energy among the beta-particles emitted in radioactive changes will fall for discussion in a later chapter; meanwhile, we may note that when that difficulty has been disposed of, energy level diagrams may be constructed (e.g. Fig. 16, p. 60) to show the relative energy content of different nuclei connected by radioactive transformations, and of different states of the same nucleus. The difference between the ordinates of any two levels may show the *kinetic* or the *total* energy evolved in transition; on the latter convention this difference is equal to the translational kinetic energy of the resultant nucleus (the parent being supposed at rest), plus the *total* energy of any particles or quanta emitted as a result of the transformation. The lowest energy state of any one nucleus is called the *ground state*; the others are *excited states.*

5. The conservation of angular momentum, and its quantization in integral and half-integral multiples of $h/2\pi$. A consistent interpretation of atomic phenomena was obtained by supposing that elementary particles such as the proton and the electron each have an intrinsic angular momentum (spin) of $\frac{1}{2}(h/2\pi)$, that their relative motions can contribute 'orbital' angular momenta that are whole multiples of $h/2\pi$, and that the radiation of a quantum involves an angular momentum change of one or more units of $h/2\pi$. Nuclear angular momenta, as deduced from (e.g.) optical spectroscopy, are found invariably to be even multiples of $\frac{1}{2}(h/2\pi)$ when the mass number of the nucleus is even, and odd multiples when it is odd; comparing this result with the quantum-theory rule for the addition of a number of equal angular momenta—namely, that the magnitude of the resultant is an even (odd) multiple of the magnitude of each component when the number of components is even (odd)—we are led to identify the mass number with the total number of particles in the nucleus. This is one of the reasons for believing nuclei to be composed of protons and neutrons rather than of protons and electrons.

6. A most important principle, deduced from the exponential decay of activity of a large assemblage of identical radioactive atoms, and consistent with the statistical fluctuations which are observed when the assemblage is not infinitely numerous, is that

the chance, per unit time, of spontaneous transition from one nuclear configuration to another depends solely upon the two states in question, and not at all upon the length of time which the nucleus has already spent in the former state. Thus each one of a number of identical nuclei has fixed probabilities $\lambda_1 dt$, $\lambda_2 dt$, . . . of spontaneous transition in time dt by the various energetically-possible processes which lead to other configurations, whether of the same or of another nuclear species. The original population will therefore decay exponentially (with statistical fluctuations) according to the equation $N = N_0 e^{-\lambda t}$, where $\lambda = \lambda_1 + \lambda_2 + \ldots$, and both the population and its rate of decrease (i.e. the activity) will fall to one-half in a time $T = \frac{1}{\lambda} \log_e 2 = 0{\cdot}693/\lambda$. T is called the half-value period or half-life, λ is the decay constant or total transition probability, and λ_1, λ_2, . . . are the partial decay constants or partial transition probabilities for the various competing processes. In practice one, or a few, of these processes will dominate the others: for a stable nucleus in its ground state all transition probabilities are, of course, zero.

The processes under discussion in this monograph may be divided into three classes; those in which the nuclear charge changes by one unit in either direction, a nuclear proton transforming to a neutron or vice versa; those in which the nucleus loses energy without changing its charge; and secondary processes external to the nucleus in question.

In terms of the constituents of the nucleus (neutrons and protons) and of the atomic number Z (equal to the number of protons in the nucleus), we may tabulate the primary processes as in table 1 (overleaf).

At least one other process is possible, namely the simultaneous creation of a positron and an electron at the expense of the energy of nuclear excitation; this occurs in some highly-excited states formed in nuclear reactions, but will not be discussed here.

The two main groupings of 'unit change in Z' and 'no change in Z' will serve as headings of later chapters; the various secondary processes will be mentioned as they arise in the course of discussion.

In spite of the absence of alpha-radioactivity, the lighter,

Table 1.

Change of nuclear constitution	Change of atomic number	Description of process	Symbol
$n \rightarrow p$	$Z \rightarrow Z + 1$	Emission of negative beta-particle (negatron) †	β^-
$p \rightarrow n$	$Z \rightarrow Z - 1$	Emission of positive beta-particle (positron)	β^+
$p \rightarrow n$	$Z \rightarrow Z - 1$	Capture of electron (usually from K shell of atom)	K
None	None	Emission of gamma radiation	γ
None	None	'Internal conversion' (ejection of electron from K, L, . . . shell of atom by transfer of energy of nuclear excitation)	e^-

artificially-prepared radioactive nuclei show more diverse modes of spontaneous change than are found among the heavy elements. The emission of a *positive* beta-particle and the capture into the nucleus of an electron from the surrounding atomic structure may on occasion combine with the familiar negative beta-particle emission to form a pattern of alternative processes more complex, if less extensive, than the typical heavy-element chains of successive alpha- and beta-disintegrations, with their rare branches. A further complication arises from the existence of nuclei for long periods of time in metastable states of internal excitation, with radioactive properties so different from those of the normal nucleus that they may be taken at first for distinct nuclear species: when recognized, such nuclei are given the convenient but non-committal name of 'isomeric forms' of the normal nuclei. Though nuclear isomerism first came to light as between two members of the uranium family (UX_2 and UZ) it is relatively much more common (or at least more frequently observed) among nuclei of medium mass.

Most radioactive changes among the lighter elements are, however, comparatively simple in themselves, consisting of a single mode of transformation either to the ground state of a stable nucleus or to an excited state from which the ground state

† This name for a negative beta-particle has recently found some favour; it certainly avoids ambiguity.

is reached almost instantaneously as a result of the radiation of one or of a few gamma-rays. As a corrective to any impression of complexity that the reader may get from later concentration upon more ' interesting ' examples, Fig. 1 illustrates the numerous but simple beta-ray transitions that involve isotopes of fluorine. The only stable nucleus is F^{19}: this can capture a free neutron

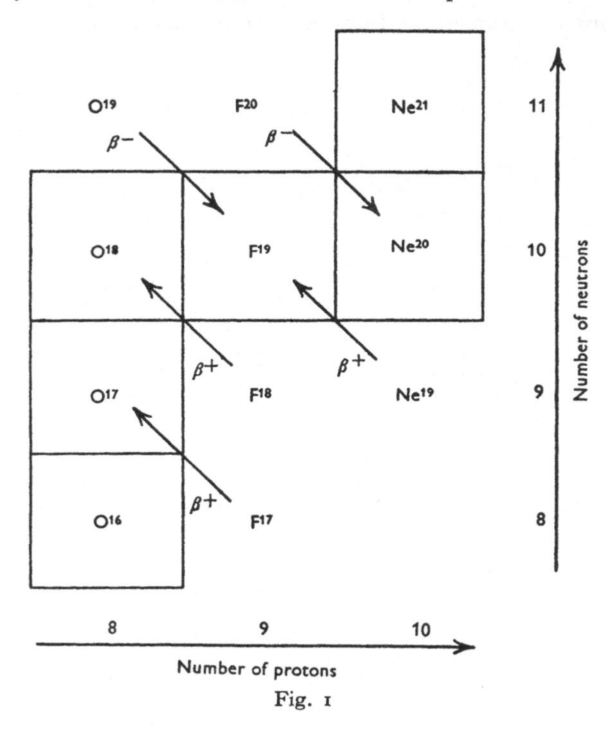

Fig. 1

with the formation of F^{20}, a nucleus which decays by negatron emission, with a half-period of 12 seconds, to the stable Ne^{20}. By ejection of a neutron from O^{17} by deuteron bombardment (a reaction symbolized by O^{17} (*d*, *n*) F^{18}), the isotope F^{18} may be formed; it decays by positron emission, of 112 minutes half-period, to the stable O^{18}. A third artificially-prepared fluorine isotope, F^{17}, decays by positron emission to O^{17}. The stable F^{19} can itself result from the positron decay of Ne^{19} or the negatron decay of O^{19}. In Fig. 1, a so-called ' neutron-proton diagram ', the nuclear constituents (numbers of neutrons and protons) serve to provide the co-ordinates of the array. The stable isotopes are enclosed in squares and the radioactive transformations are shown by arrows.

In such a diagram, *isotopes* (nuclei of equal charge-number Z) lie in the same vertical column; *isobars* (nuclei of equal mass-number A) lie along a diagonal of negative gradient 45°. Beta-transitions are transitions between neighbouring isobars.

CHAPTER II

IDENTIFICATION AND MEASUREMENT OF RADIATIONS AND PARTICLES

The study of radioactive nuclei is closely bound up with the techniques of detecting and measuring the energies of the particles and radiations they emit. To avoid frequent explanations later, the main experimental methods are listed and briefly discussed now. For thorough discussions and technical details, the references given in, and at the end of, this chapter should be consulted.

1. Beta-particles and Electrons (β^-, β^+, e^-)

A. Detection

These particles may be detected by the Geiger-Müller counter, by the electron multiplier (using the principle of multiplication by secondary electron emission, in successive stages), by the Wilson cloud chamber, by an ionization chamber and an electroscope or its equivalent, or by the blackening of a photographic emulsion. The first three methods permit the detection of single particles, as do recent special photographic emulsions.

A counter will detect electrons with virtually 100% efficiency, provided that its walls are thin enough for the particles to enter; it can register the time of passage of a particle with high accuracy, so that two counters can determine whether or not two particles discharging them are simultaneous. The same principle can be extended to any number of particles. The usual circuit for detecting such 'coincidences', due to Rossi, is indicated in Fig. 2. Only when negative impulses I_1, I_2, . . . I_n make *all* the pentodes simultaneously non-conducting will the common anode voltage rise to anything like the full battery voltage. The present limit of resolution of Geiger counter coincidences approaches $\frac{1}{10}\mu$ sec., though it is likely to be capable of further improvement (see, e.g., *PRS*, 177, 260; *PR*, 59, 706; *HPA*, 16, 251).

The electron multiplier (e.g. *RSI*, 12, 484; *HPA*, 19, 211), though it will not detect every electron that strikes its first cathode, is capable of much higher resolution of coincidences, provided that the whole amplifying circuit, up to the point where coincidence or non-coincidence between the pulses from the two multipliers is assessed, has sufficiently rapid response.

In the field of artificial radioactivity, the cloud chamber is of most use for particles emitted either from gas atoms in the chamber or from sources on the surface of solid bodies placed within the chamber. It enables the point of origin, the range, the density of ionization along the track and the curvature of

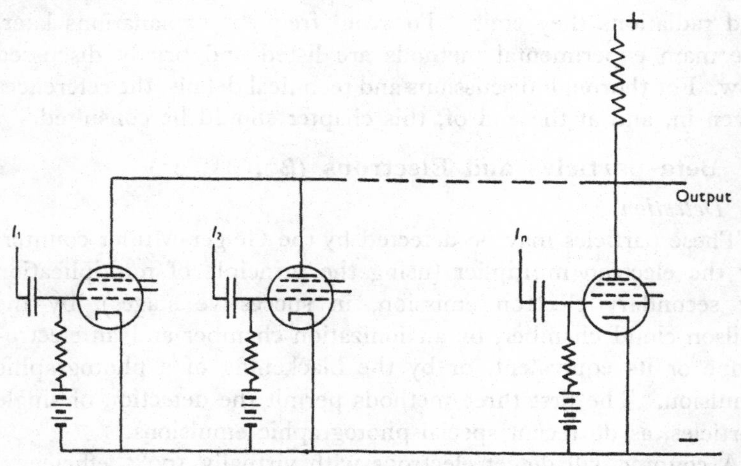

Fig. 2.—Coincidence circuit: negative voltage impulses are injected at points $I_1, I_2, \ldots I_n$.

the track in a magnetic field to be determined for each particle observed. The efficiency of detection of particles arriving just before the expansion is 100%. Tracks arriving too early will be lost by the diffusion and recombination of the ions they produce; the time required for this to happen depends upon various factors, but is often of the order of a tenth of a second. The cloud chamber is particularly useful for detecting slow electrons (say, below 50 keV.), especially those which are emitted from gaseous atoms.

The ionization chamber is important for the accurate measurement of activities and decay periods for substances obtainable in relatively large quantity, when the rate of arrival of particles is so high that it is unnecessary to count single electrons. A sensitive and convenient apparatus of this type is the Lauritsen electroscope (*RSI*, 8, 348).

The blackening of a photographic plate by electrons is rather insensitive and is principally used in β-ray spectographs, where this disadvantage is offset by its property of integrating over long exposures and by its ability to register a whole spectrum at once.†

Positrons and electrons are distinguished by the sense in which they are deflected in a magnetic field—frequently the field of a spectrograph in which their momenta are being investigated. Confirmation of positron emission may be obtained by observing the quanta (of 0·51 MeV. energy) emitted in pairs in opposite directions when the positrons, after being reduced to rest by collisions in whatever material surrounds the source, are annihilated together with electrons.

B. Energy measurement

The energies with which electrons escape from the source can be estimated by measuring their absorption in air or, more conveniently, in foils of a light material such as aluminium. Absorption measurements are less satisfactory for electrons than for α-particles, or other relatively heavy particles such as protons or deuterons, which have very definite ranges in matter. A beam of electrons, initially well-collimated and homogeneous in energy, rapidly loses homogeneity of energy and direction as it passes through matter. Nevertheless, a considerable amount of information has been won by careful absorption measurements. For example, the upper limit of a continuous β-ray spectrum can be found with good accuracy by comparing the shape of the end-portion of its absorption curve with that of a 'standard' substance of known end-point (*PCPS*, 34, 599; *HPA*, 19, 375). Feather's formula relating the energy E (MeV.) to the extreme range R (g./cm.^{2}) of the β-particle is

$$R = 0{\cdot}543E - 0{\cdot}160,$$

and is a useful approximation to the result of such detailed comparison.

Conversion lines superposed upon a continuous β-ray spectrum can sometimes be identified by kinks in the absorption curve.

† The counting of individual electron tracks in a special emulsion is too difficult.

The change of slope at each kink is small, and the measurements must be made very carefully if the method is to succeed (see, e.g., *AP*, 19, 219).

For accurate and reliable energy values, measurements are usually made in terms of the curvature of path of the particle in a magnetic field of known strength. Owing to the continual loss of velocity and the small random changes of direction that arise from collisions with gas molecules, measurement of the curvature of a β-particle track in a cloud chamber in a magnetic field gives only moderate accuracy, and the use of a vacuum β-ray spectroscope is preferred whenever possible.

β-ray spectroscopes differ in the manner in which particles of a given momentum are brought to a point or line focus when their initial source-points and directions of emergence differ slightly. The types hitherto most used are as follows:

(a) *Semicircular focusing* (Danysz), in which a narrow pencil is selected from the rays leaving a point (in practice, a short line source) in directions nearly perpendicular to the uniform magnetic field, particles of given momentum coming to a line focus (with a sharp outer edge). The path from source to focus is a semicircle for the central ray, a little more or less than a semicircle for rays deviating slightly from this in the plane perpendicular to the field, and a flat helix for rays slightly inclined to this plane. With E in MeV., $B\rho$ in gauss. cm.,

$$E = 0{\cdot}511(\sqrt{1 + (B\rho/1704)^2} - 1).$$

(b) *Helical focusing* (Tricker), in which the source is a small disk lying perpendicular to the uniform field, and particles leaving the source at small angles with the field direction are brought to an axial focus after describing a complete turn of an elongated helix. This type is now obsolescent.

(c) *Magnetic lens spectrographs*, in which the system of focusing is, as in the Tricker method, axially symmetrical, but where the paths are controlled by the more or less localized field of a magnetic lens. By careful shaping of the field it is possible, with a given degree of resolution, to accept particles leaving the source over a larger range of directions than the other systems will allow, but it has not yet been found possible to accept the same small solid angle that is used in a high-resolution semicircular system, and secure higher resolution than that system affords.

A potentially important modification of the 'semicircular' method, in which a magnetic field varying as $1/r^{\frac{1}{2}}$ gives convergence of a conically-divergent beam to a point focus at $\pi\sqrt{2}$ radians, has been suggested by Siegbahn and Svartholm (*N*, 157, 872; see also *PR*, 71, 681 and 72, 256).

At the moment, then, the semicircular method is usually employed when it is important to obtain either absolute values of the momentum (and hence the energy), or the best possible resolution. The magnetic lens type is most attractive when calibration with particles of known momentum is possible and when high intensity is important. It may not be long before the magnetic lens spectrograph competes with the semicircular as regards attainable resolution, but, since only a uniform field can be measured in absolute terms with the highest possible accuracy, the semicircular type is likely to remain the basis of absolute measurements of energy.

With any of the above systems, the particles may be detected either by a photographic plate or by a counter, the plate indicating in a single exposure the positions of lines over a considerable range of the spectrum, the counter giving reliable measurement of relative intensities of the lines or of different parts of a continuous spectrum, provided of course that loss of energy in the source is negligible, and that the counter can admit the least energetic particles which it is desired to investigate. The semicircular type of spectrograph with its line focus is well suited for use with a cylindrical counter, entry of particles being limited by a slit parallel to the axis of the tube. Bell-shaped counters with circular apertures (*RSI*, 14, 205) are suitable for lens or helical spectrographs.

It should be noticed that the variation of blackening along a spectrum recorded by a photographic plate does not immediately give the relative numbers of particles of different momenta, since the blackening 'per electron' depends greatly on the energy of the electrons striking each portion of the plate. When a counter is used, it is necessary not only to be sure that the window is thin enough for all the particles concerned to penetrate into the counter, but also that the pressure of gas in the counter is such that even the most energetic particles are certain to produce ion pairs in passing across the counter. In air at a pressure of 10 cm. of mercury, a 1 MeV. electron produces only about 3 ion pairs per centimetre.

If a continuous β-ray spectrum is being investigated by varying the strength of the deflecting field and keeping the counter in a fixed position, particles over a fixed *percentage* range of momentum will enter the counter. The rate of counting must therefore be divided by the field strength in order to obtain a correct momentum-distribution curve. Such a correction is not required when comparing the intensities of lines, for then it is a question of the total number of particles in the group, and not the number per unit range of momentum.

Spectroscopes employing electrostatic fields are scarcely practicable for β-rays of average energy, since the voltages required would be very high, but an electrostatic spectroscope has been used by Backus to investigate the lowest energy positrons and negatrons emitted by Cu^{64}. (See p. 62.) Watts and Williams (*PR*, 70, 640) have used the simplest of all systems to estimate the upper limit of the β-ray spectrum of H^3 (about 11 keV.); the source was placed opposite to the thin window of a counter, and accelerating or retarding potentials were applied between source and counter until the β-rays just ceased to penetrate the window, the stopping power of which was determined by a similar experiment with electrons accelerated from a hot filament.†

C. *Measurement of absolute strengths of β-ray sources*

The number of β-particles leaving a source per second is obviously best measured by actually counting particles rather than by measuring ionization or by using a photographic plate. The problem is then to determine the efficiency of the counter in detecting the particles concerned.

We have so far used ‘ efficiency ’ in the rather vague sense of the chance that a particle or photon passing through the counter shall be detected and registered. As soon as we consider the matter more closely we see that it is bound to depend on the orientation of the counter with respect to the source, on whether the beam of particles is parallel or divergent, and so on. It has

† An upper limit of 17 keV. has since been found by another method (*N*, 162, 302).

been emphasized by Dunworth (*RSI*, 11, 167) that the quantity that is of practical importance and can be precisely defined is the so-called 'net efficiency', which is the chance that a particle or quantum emitted in a random direction by an atom of the source shall be detected by a given counter in a given position with respect to the source. Methods of measuring this quantity usually involve the counting of gamma-rays as well as beta-particles and will be described below.

2. Gamma-rays

Gamma-rays of moderate energy are detected by way of the photoelectrons and Compton recoil electrons which they eject from atoms among which they pass. Fig. 3 shows the way in which the probability of these two processes varies with the energy of the photon and with the atomic number of the atoms. The average distance which a gamma-ray travels through matter before producing an electron is large compared with the range of the electron in the same material, so the efficiency of detection by a cloud chamber or counter of reasonable volume, or by a photographic emulsion, is quite low. In a counter, since most of the electrons come from the walls, the best that can be done is to make the cathode of an element of high atomic number, which gives more photoelectrons. Platinum and bismuth are probably the best, since lead is apt to be contaminated with heavy radioactive elements. The cathode may be in the form of a wire mesh to increase the surface area. Even so, the intrinsic efficiency will be of the order of only 1%, and the net efficiency will be perhaps a tenth of this, because the solid angle subtended by the counter at the source will usually be of the order of a tenth of 4π, even in favourable arrangements.

An important recent development is the use of photomultiplier tubes to detect the scintillations produced by γ-rays in certain crystals, such as naphthalene, anthracene and calcium tungstate. If a thick transparent crystal can be used, each γ-ray passing through it has a high chance of Compton scattering or (if heavy atoms are present) photoelectric absorption; the many photons generated by each photo- or Compton electron go out in all directions and there is a high chance that enough of them will strike the photosensitive cathode to release several electrons from it. The intrinsic efficiency of the system as a γ-ray counter can thus approach unity, and the resolving time can be extremely short.

These same two processes of photoelectric effect and Compton scattering form the foundation of the most usual methods of measuring the quantum energies of γ-rays, the energies of the Compton- or photo-electrons from a suitable 'radiator' through which γ-rays pass being measured with a magnetic spectroscope

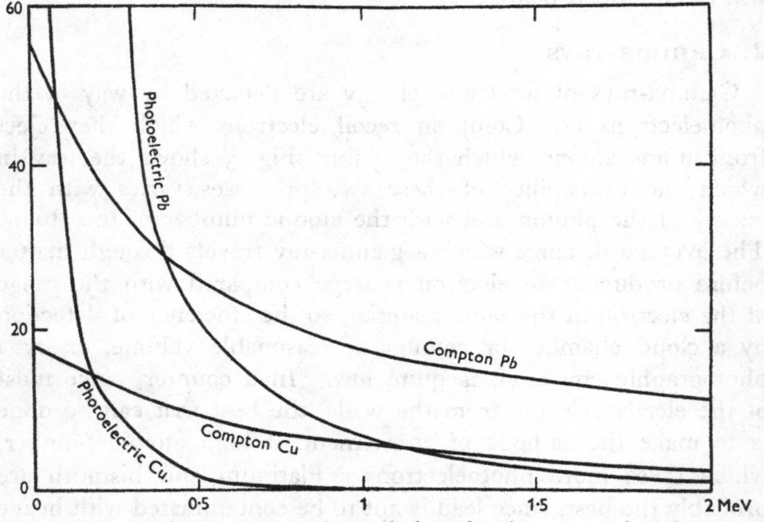

Fig. 3.—Compton scattering and photoelectric cross-sections in lead and in copper (units 10^{-24} cm.^{2} *per atom*).

or in a cloud chamber, or by their absorption end-points (*PRS*, 169, 269), best taken in coincidence (p. 22) with parent betas.

The relative probability of the two processes is seen from Fig. 3 to depend strongly upon the atomic number of the radiator material and the γ-ray energy. If Compton electrons only are desired the radiator should clearly consist of an element of low atomic number. The energy of the ejected electrons does not depend appreciably upon the atomic number, since the energy required to remove an electron from a light atom (or indeed from any but the few innermost levels of a heavy atom) is small compared with the energy which may be imparted to the electron by a reasonably energetic γ-ray; the Compton scattering process is essentially that of the elastic collision of a γ-ray with a free electron. The energy imparted to the electron depends upon the direction (with respect to that of the incident quantum) in which the electron is projected, and is greatest in the forward direction. The ideal arrangement would therefore be for the gamma-rays to be in a roughly unidirectional beam, for only those electrons to be accepted which leave the radiator in that same direction, and for the radiator to be thin compared with the range of the Compton electrons.

Unfortunately, these conditions can seldom be fulfilled, for the following reasons. Firstly, considerations of intensity usually

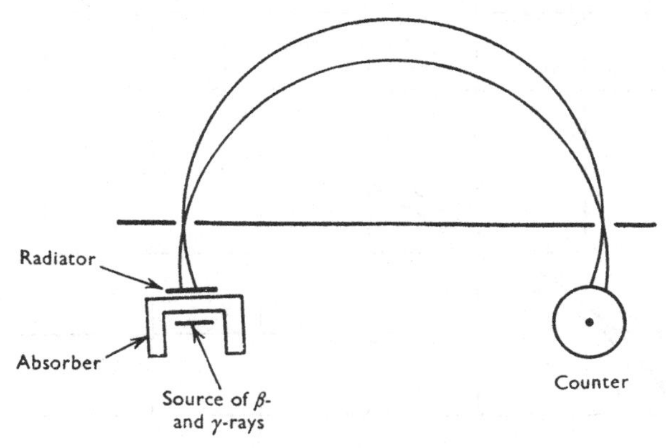

Fig. 4.—Arrangement for measuring energies of γ-rays from β-γ source.

require the active specimen to be placed quite close to the radiator, which is therefore bombarded with γ-rays from a wide range of directions. Secondly, the specimen will usually emit its own β-rays, which are far more numerous than the Compton electrons and must be prevented from entering the spectroscope. An appropriately thick absorber must therefore be placed between the source and the radiator, and since Compton electrons will be generated at all depths within this absorber there is little or nothing to be gained by having a separate radiator.

It is therefore usual to accept the practical limitations of the method, to place over the source an absorber of a light element (often Al or Cu), thick enough to eliminate β-rays or conversion electrons from the source, and to measure the energy distribution of the Compton electrons which emerge from the absorber (Fig. 4). Each γ-ray line gives a broad ' hump ' in the distribution curve, the upper energy limit of the hump † corresponding to electrons projected in the forward direction from the surface of the absorber. In this way unambiguous if not highly accurate values for the γ-ray energies can be obtained. In terms of the radius of curvature ρ of the path of these forward-projected electrons supposed moving at right angles to the magnetic field

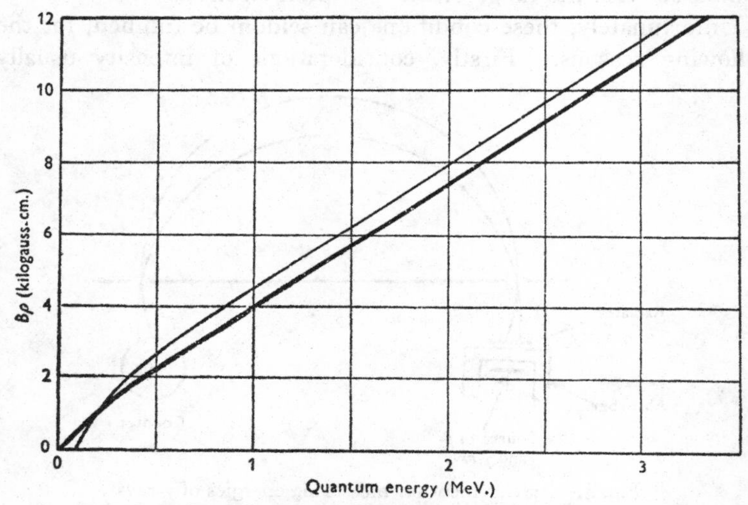

Fig. 5.—Values of $B\rho$ for forward-projected Compton electrons (heavy line) and for K-shell photoelectrons from Pb (light line).

of induction B, the energy of the *photon* in practical units is

$$E = h\nu = 0{\cdot}255\left[\frac{B\rho}{1704} + \sqrt{1 + \left(\frac{B\rho}{1704}\right)^2} - 1\right] \text{MeV}.$$

Fig. 5 shows a graph of $h\nu$ against $B\rho$.

If a thin radiator of *high* atomic number is now placed over the absorber, each γ-ray line of energy $h\nu$ will give monoenergetic photoelectron groups of energies $h\nu - E_K$, $h\nu - E_{L_I}$, . . . , where E_K, E_{L_I} . . . are the energies required to remove a K, L_I, . . . electron from an atom of the radiator and are well known from X-ray experiments. Thus sharp lines will appear superimposed upon the broad Compton electron spectrum: their

† Account may have to be taken of instrumental spreading of the line (see *AMAF*, 30A, No. 1).

assignment to K, L_{I}, . . . levels is greatly aided by the information provided by the Compton spectrum. For an example, see p. 83.

In the many instances where the energy of the nuclear transition can be communicated to one of the inner electrons of the source atom in question, instead of being radiated as a quantum, groups of 'internal conversion' electrons will appear in the magnetic spectrum of the bare source: their energies are equal to $h\nu - E_K$, $h\nu - E_{L_I}$, etc., where E_K, E_{L_I} . . . are the binding energies of the K, L_{I}, . . . electrons of the atom in which the internal conversion takes place. If the conversion electrons follow a radioactive change, the atom in question is the ***product*** of that change.

When the substance emits negatrons, the internal conversion lines are superposed on the β^--ray spectrum: when positrons are emitted, the β^+-particles and the conversion electrons will of course be observed in the spectroscope with oppositely directed magnetic fields, but the electron spectrum will contain a line-component corresponding to the γ-rays of 0·51 MeV. emitted at the annihilation of the positrons in whatever material they are brought to rest.

It will be noted that the β-ray spectrum, the internal conversion electron spectrum, the Compton spectrum, and the external photoelectron spectrum may be investigated with the same arrangement of source and spectroscope: it is only necessary to cover the source with an absorber of a light element in order to replace the β-ray and internal conversion spectrum by the Compton spectrum, and to cover this in turn with a thin radiator of high atomic number in order to add the photoelectron spectrum.

When, as quite often happens, an internal conversion line but no corresponding Compton group or external photoelectron line is observed, that particular transition is presumably going entirely by conversion.

Pair creation

A γ-ray of energy greater than about 1 MeV. can create a positron and an electron in the neighbourhood of a nucleus, which serves to take up momentum but, because of its large mass, acquires only a negligible amount of recoil energy. Energy equal to mc^2 is required to create each particle of rest-mass m, and the remaining energy $h\nu - 2mc^2$ is shared, usually unequally, as kinetic energy between them.

By coincidence methods it is possible to discriminate between pair production and other processes such as the Compton effect, which yield only one particle at a time, but unless the kinetic energy of *both* particles is measured the energy of the γ-ray cannot be deduced. The cloud chamber is the only apparatus that enables this to be done efficiently.† A pair of tracks starting from a single point is easy to identify, particularly with the aid of a magnetic field, when the characteristic shape shown in Fig. 6 is obtained. The particles will usually leave the illuminated region

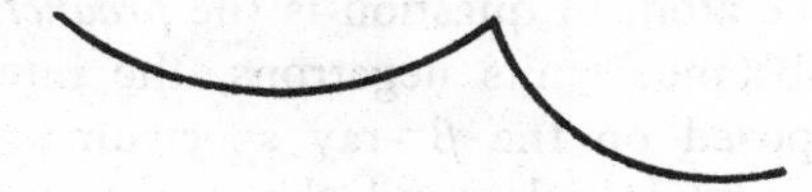

Fig. 6.—Typical geometry of cloud-chamber tracks of positron - electron pair.

of the chamber before they have travelled any large fraction of their total range, but the curvature may increase perceptibly along the visible portion of each track owing to the progressive loss of energy through ionization. Careful curvature measurements of 'stereoscopic' photographs taken simultaneously from two roughly perpendicular directions can give accurate values of the initial energy of each particle; the energy of the γ-ray is obtained by adding $2mc^2$ (= 1·02 MeV.) to the sum of these energies (*PR*, 67, 273).

Nuclear photo-effect

The nuclear photo-effect (the ejection of a neutron from a nucleus following the absorption of a γ-ray) is usually applied to γ-ray energy measurements through the reactions

$$D_1^2 + h\nu = H_1^1 + n_0^1 \quad \text{and}$$

$$Be_4^9 + h\nu = Be_4^8 + n_0^1,$$

the threshold energies for which are 2·18 and 1·62 MeV. respectively.

† For γ-rays above about 3 MeV., a 'pair spectroscope' using counters is practicable (*PR*, 74, 315).

The yields are small (cross section 6×10^{-28} cm.^{2} for the former reaction when $h\nu = 2{\cdot}6$ MeV.), but even a small yield of neutrons is readily identified in the presence of a much larger number of any other type of particle, and so γ-rays of energy above the threshold are detectable with certainty in the presence of softer γ-rays or of β-particles.

The energy of the γ-ray can be inferred either from the kinetic energy of the photoneutron or from the ratio of the numbers of photoneutrons ejected from D and from Be. A better way is to measure photo*proton* energies in a deuterium ionisation-chamber.

Absorption measurements

Gamma-ray energies can be estimated by absorption methods: that is, by interposing different thicknesses of suitable elements between the source and whatever detector is used. This assumes the absorption coefficient to be already known as a function of gamma-ray energy. Several difficulties are encountered: Compton scattering as well as true absorption is present, so the decrease of measured intensity depends on the geometry, while at very high γ-ray energies the absorption coefficient increases again owing to the onset of pair-production, so two different energies have the same absorption coefficient in a given material. This difficulty can be resolved by measuring the absorption in a light element (Al) for which the minimum absorption occurs at an energy well beyond the greatest γ-ray quantum energy with which we are likely to be concerned. At low energies another type of ambiguity occurs owing to the sudden change of photoelectric absorption at the K, L . . . . X-ray absorption limits, but the suddenness and magnitude of this change (often a factor of three or more) makes it a valuable means of estimating the energies of soft γ-rays by comparing their absorption in neighbouring elements.

Crystal spectroscopes

The quantum energy $h\nu$ of a γ-ray can of course be immediately calculated from its wavelength c/ν, and this wavelength can in principle be measured by the methods of X-ray spectroscopy. In practice, this is difficult, not only on account of the very small angular deflections, but also because of problems of intensity.

The classical work in this field is that of Rutherford and Andrade (*PM*, 27, 854; 28, 263), who investigated the low-energy γ-ray and X-ray lines of the 'active deposit' (RaB, RaC, RaC′, RaC″) formed by the decay of radon. A source of the order of 100 millicuries gave photographic records of reflexions from a rock-salt crystal with exposures of 24 hours; the shortest wavelength measured was $0{\cdot}071 \times 10^{-8}$ cm., corresponding to an energy of 0·175 MeV. The glancing angle was about 44′.

Gamma-ray lines of energies as high as 0·77 MeV. have since been observed with a crystal spectroscope, but the method has seldom been used to investigate nuclear gamma-rays, which mostly have energies beyond its range. Curved-crystal spectroscopes, which give higher intensities, have recently been used for studies of the X-rays that follow internal-conversion or electron-capture processes (*PR*, 69, 140), and for precise gamma-ray measurements (*PR*, 73, 1392).

Gamma-ray branches and cascades; applications of coincidence methods

Each gamma-ray results from the transition of a nucleus from one to another of its many energy levels, and a nucleus in any given excited state will have a finite probability of transition to each lower state: if the two or more strongest competitors are of similar orders of probability, some of the identical nuclei in that given excited state will emit one gamma-ray and some another: these γ-rays are ***in competition*** and the nucleus shows ***γ-ray branching***. The branching ratio will be a constant for that particular excited state of the nucleus, however it was formed. Branching can often be identified by energy relationships between the γ-ray lines: for example, if a nucleus gives γ-rays of energies 1·20, 0·80 and 0·40 MeV., it is likely that the first transition is an alternative to the two others, which occur in succession (or 'cascade') via a level of intermediate energy.

In the example given, energy measurements will not by themselves determine which of the cascade γ-rays is emitted first, and so the level scheme might be either that of Fig. 7 (*a*) or of Fig. 7 (*b*); this question can be settled if it is possible to produce the excited nucleus by some other reaction which happens to give it only low excitation energy and thus to produce only that line of the γ-ray spectrum corresponding to transition between the two lowest levels.

The elucidation of γ-ray spectra has been greatly helped by coincidence counting methods. Gamma-rays which are in competition are emitted by different individual nuclei and coincide in time only by chance, while cascade γ-rays, emitted in succession from the same nucleus at intervals of time usually far below the limit of instrumental resolution, will cause coincident discharges between two or more counters. Again, a β-ray change leading to an excited state of the product nucleus will be accompanied by γ-rays of that nucleus, and by counting only those γ-rays

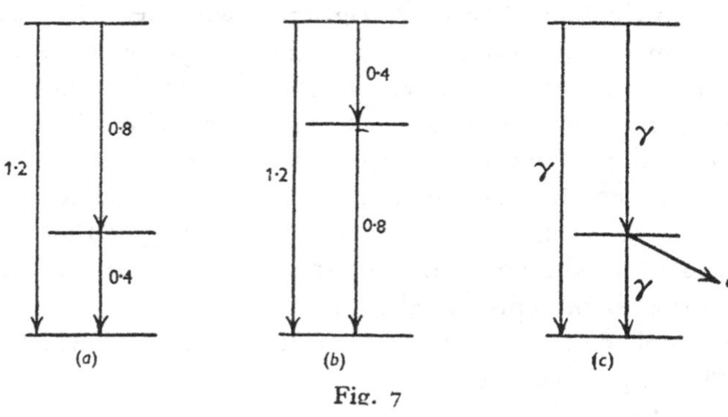

Fig. 7

which are coincident with β-rays we may largely eliminate counts due to cosmic rays, or to any other process which is not an instantaneous result of the β-ray change.

If alternative β-ray transitions lead to the ground state and an excited state of the product nucleus, the γ-rays will show coincidences only with the lower-energy β-rays, and the complex nature of the β-ray spectrum is thereby demonstrated. The β-rays may be analysed either by absorbers or by a spectrograph; with the former arrangement (Norling, Feather) an absorption curve for the low-energy partial spectrum is obtained, with the latter, coincidences may be observed between β-rays *of given energy* and the accompanying γ-rays.

Coincidence methods may be used in conjunction with γ-ray absorption measurements: if the rate of coincidence between two γ-counters is measured as a function of the thickness of absorber placed between the source and one of them, the absorption curve can be analysed approximately into exponentials, each of which represents one of the various γ-rays that are emitted simultaneously (from the point of view of the counters) and therefore are actually emitted in cascade from one and the same nucleus. Thus the energies of members of a cascade can be estimated from their absorption coefficients, γ-rays which are not members of the cascade being eliminated because they give only accidental coincidences.

Downing, Deutsch and Roberts (*PR*, 61, 686) point out that if one member of the cascade has a lower probability than another (as in Fig. 7 (*c*), where one γ-ray is partly forestalled by internal conversion) the *coincidences* between them will vary with absorber thickness in just the same way as if the two γ-rays were equally numerous. This can be shown as follows. Let N nuclei radiate the first γ-ray per second, and let the chance of its detection by either of two counters, assumed identical, be $\mathcal{E}_1$. Let the fraction of nuclei which, having radiated the first γ-ray, radiate also the second, be α, let the chance of its detection by either counter be $\mathcal{E}_2$; also let the absorption of the two γ-rays by thickness x of the absorber be represented by $e^{-\lambda_1 x}$, $e^{-\lambda_2 x}$.

The number of coincidences recorded per second will then be $N\mathcal{E}_1 e^{-\lambda_1 x} . \alpha\mathcal{E}_2 + N\mathcal{E}_1 . \alpha e^{-\lambda_2 x}\mathcal{E}_2$.

The first term is (rate of counting of γ-ray '1' through the absorber) $\times$ (chance that γ-ray '2' is emitted and is detected by the other counter), while the second term is (rate of counting of γ-ray '1' by the counter which has no absorber) $\times$ (chance that γ-ray '2' is emitted and is detected by the counter behind the absorber). The sum of the two terms is $N\mathcal{E}_1\mathcal{E}_2\alpha(e^{-\lambda_1 x} + e^{-\lambda_2 x})$ and, as stated above, varies with x in a manner independent of α. If equal absorbers had been placed in front of *both* counters, the coincidence rate would have been $2N\mathcal{E}_1\mathcal{E}_2\alpha e^{-(\lambda_1 + \lambda_2)x}$, and only the sum of the two absorption coefficients would have been obtained.

Another important method (Dunworth, *RSI*, 11, 267) employs $\beta\gamma$ and $\gamma\gamma$ coincidences and is applicable to cases where something is already known about the disintegration scheme: it may be explained by a simplified example. Suppose a radioactive specimen is known to emit β-rays each followed by two γ-rays in cascade. Let the unknown strength of the source be N disintegrations per second and the net efficiency of each of a pair of identical and symmetrically placed γ-ray counters be $\mathcal{E}_1$, $\mathcal{E}_2$ for the gamma-rays '1' and '2'. The rate of single counting of either counter is $N(\mathcal{E}_1 + \mathcal{E}_2)$, while the rate of coincidence counting is $N\mathcal{E}_1 . \mathcal{E}_2 + N\mathcal{E}_2 . \mathcal{E}_1 = 2N\mathcal{E}_1\mathcal{E}_2$. Let one of the γ-ray counters be replaced by a β-ray counter of net efficiency $\mathcal{E}_3$. The single counting rate of the β-ray counter is $N\mathcal{E}_3$, while the $\beta\gamma$ coincidence rate is $N\mathcal{E}_1\mathcal{E}_3 + N\mathcal{E}_2\mathcal{E}_3 = N\mathcal{E}_3(\mathcal{E}_1 + \mathcal{E}_2)$. The four measured rates suffice for calculating the four unknowns N, $\mathcal{E}_1$, $\mathcal{E}_2$, $\mathcal{E}_3$. The equations are little more complicated, and equally well soluble, if the two γ counters are not identical. Clearly we have here a principle of great power and elegance: in practice, considerable judgment must be exercised to compromise between (1) the predominance of cosmic ray coincidences and the low rate of accumulation of information when the source is weak, and (2) the excessive number of *accidental* $\beta\gamma$ and $\gamma\gamma$ coincidences when the source is strong. These matters are fully discussed by Dunworth.

Once the net efficiency of a given γ-ray counter (in a given position with respect to the source) has been determined for a number of γ-rays of previously known energy, the argument may be reversed and the energy of a γ-ray deduced from the net efficiency with which the counter detects it.

By dividing the observed net efficiency of a counting arrangement by the fractional solid angle subtended by the counter at an average point of the source, one may estimate the intrinsic efficiency of the counter; that is, the probability that a quantum striking the counter shall be registered. As has already been mentioned, this probability will depend somewhat upon the place and angle at which the photon enters the counter, and the intrinsic efficiency is, to the actual experimenter, a vague as well as an unimportant quantity.

To anybody else, however, the net efficiency gives no impression of the performance of the counter itself, so to illustrate the way in which the sensitivity of gamma-ray counters depends upon the material of the counter wall and upon the energy of the gamma-rays, *intrinsic* efficiencies as given by Bradt and his colleagues (*HPA*, 19, 77) are plotted in Fig. 8. It will be seen that over most of the range the efficiency increases with energy, the increasing depth from which photo- and Compton electrons emerge outweighing the decreasing probability of occurrence of these two processes; the Pb counter has relatively better efficiency at low energies because of the higher yield of photoelectrons from a material of high atomic member. The intrinsic efficiency of even the best γ-ray counter being low, experiments involving γ-ray counting (and particularly $\gamma\gamma$ coincidences) are apt to be difficult. Photomultiplier counters (p. 15) may change this.

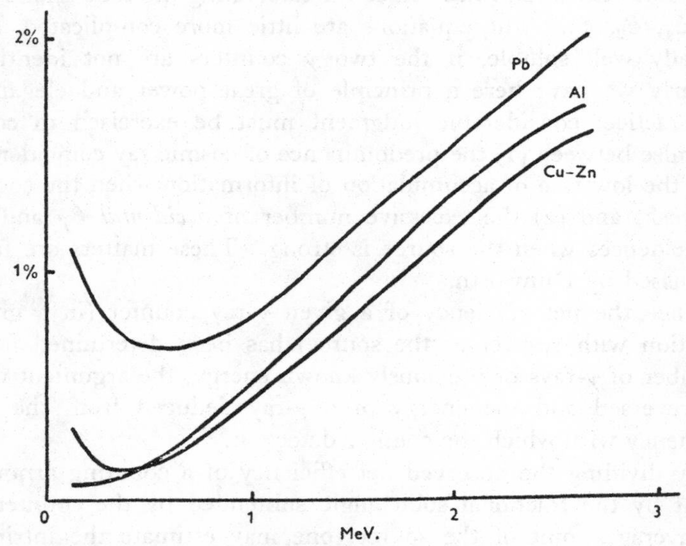

Fig. 8.—Intrinsic efficiencies of γ-ray counters. (Bradt *et. al.*)

In explaining the coincidence method of determining net efficiencies, it was assumed that every β disintegration gives one of each of the two γ-rays; if this is not true the apparent values of gamma-ray efficiency will fall below the curve, thereby indicating the presence of internal conversion or γ-ray branching. Progress in analysing $\beta - \gamma$ processes can be made most quickly and certainly by the combination of coincidence experiments with β- and γ-ray spectroscopy; examples will be found in later chapters.

3. X-rays.

Characteristic X-rays are emitted whenever a vacancy occurs in the K (or any inner) shell of an atom. The identification of

X-rays, and of the element by which they are radiated, is often important as a means of identifying the primary 'radioactive' phenomena (electron capture, and internal conversion) which can cause vacancies in the inner electronic shells of the atom. X-rays are of course physically identical with low-energy γ-rays † and the same general methods may be used for their detection: energy measurements for the purpose of identifying the radiating atom are usually made by the 'absorption limit' method mentioned on p. 21 in connexion with soft γ-rays. In energy, the main K radiation of any element falls between the K absorption limits of two neighbouring elements of slightly lower atomic numbers; for example, the K radiation of zinc ($Z = 30$) has a quantum energy lying between the energies of the K-absorption limits of copper ($Z = 29$) and nickel ($Z = 28$), and is therefore much more strongly absorbed in nickel than in copper. The element which is the source of any observed K or L line can generally be identified so simply and certainly by this 'critical absorption' method that the more elaborate methods of X-ray spectroscopy are rarely used.

4. Identification of Radioactive Isotopes.

Several methods can help to identify the element, and the isotope of that element, which is responsible for any observed radioactive change; but it is usually necessary to rely on the fitting together of several pieces of evidence to obtain a positive identification.

The chemical nature of the active isotope can be found (if the decay period is not too short or the chemical separation from neighbouring elements not too difficult, as with the rare earths) by mixing, in ionic solution, the active material with small amounts of the ordinary inactive forms of the suspected elements, and separating these elements from one another by appropriate chemical or physical means. The atoms of the active material will be 'carried' with the inactive atoms with which they are isotopic. The carrier is necessary or, at the least, desirable, to provide 'bulk' for the formation of filterable precipitates, to prevent appreciable loss of active material by adsorption on the walls of the vessel, and so forth; in short, to ensure that the extremely

† The K lines of uranium have quantum energies of about 0·10 MeV.

small quantity of the radioactive element has the ordinary analytical behaviour of that element

The actual isotope responsible for the activity can sometimes be decided with tolerable certainty from a knowledge of the stable isotopes in the region of the periodic table concerned, and from experience of the usual result of the particular nuclear bombardment by which the activity was produced. For example, an activity produced by the action of slow neutrons upon iodine, which has the single stable isotope I^{127}, will almost certainly be due to neutron capture and will be I^{128}. If a different activity, chemically identified as tellurium, were produced by the action of *fast* neutrons upon iodine, it would very probably be Te^{127} resulting from an (n, p) reaction.

In general, however, it is necessary to compare the half-periods and other characteristics of radioactivities produced by various nuclear reactions before an assignment can be made; as a simple example we may consider the activities resulting from the bombardment of silver (stable isotopes 107 and 109) by (a) slow neutrons and (b) energetic gamma-rays.

The slow neutron bombardment gives two β^--bodies of half-periods 22 seconds and 2·3 minutes; these will be Ag^{108} and Ag^{110}, but we do not yet know which is which. Irradiation of silver with hard gamma-rays yields the same 2·3-minute activity together with a 25-minute positron activity. These must be produced by the *removal* of a neutron from the two stable isotopes. Clearly then the 2·3-minute activity is due to Ag^{108}.

Valuable confirmation of such assignments can nowadays be obtained by artificial separation of isotopes, either of the target material or of the radioactive products.

General References

" Report on the cloud chamber and its applications in nuclear physics." N. N. das Gupta and S. K. Ghosh, *Reviews of Modern Physics*, Vol. 18, p. 225 (April, 1946).

Electron and nuclear counters. S. A. Korff (Van Nostrand: New York, 1946).

" Coincidence methods in nuclear physics." J. W. Dunworth, *Review of Scientific Instruments*, Vol. 11, p. 167, 1940.

Electrical Counting. W. B. Lewis (Cambridge University Press, 1942).

"Studies in β-ray spectroscopy." K. Siegbahn, *Ark. Mat. Ast. Fys.*, 30A, No. 20 (Stockholm, 1944).

"The use of coincidence counting methods in determining nuclear disintegration schemes." A. C. G. Mitchell, *Reviews of Modern Physics*, Vol. 20, p. 296 (January, 1948).

CHAPTER III

RADIOACTIVE PROCESSES IN WHICH Z CHANGES BY ± 1

Three processes are experimentally observed by which a nucleus can spontaneously change its charge by one unit, namely,

(1) the emission of a negative beta-particle (negatron),

(2) the emission of a positive beta-particle (positron),

(3) the capture of an electron from the K-shell of the atom.

The first two processes are usually detected by way of the β-particles themselves, but electron capture can be observed only through some secondary phenomenon—often the X-radiations which are emitted when the vacancy in the K-shell is filled by an electron from the L-shell, the resulting L vacancy from the M-shell, and so on.

Beta-particles are never emitted as the ' immediate ' consequence of a nuclear reaction; in the high state of nuclear excitation which follows a nuclear collision, the effective competitors are the much more rapid processes of heavy particle emission and gamma-radiation: when, as a result of one or both of these processes, the nucleus has lost so much energy that it can no longer emit a heavy particle, gamma-radiation is normally the dominant process, and only when this is inhibited by the phenomenon of metastability can β^-- or β^+-emission, or electron capture, occur in an excited nucleus.

Thus the three processes which we are now considering usually start from the *ground state* of a nucleus, though, as we shall see, they often lead to an excited state of the product nucleus.

The most striking experimental feature of the beta-decay process is that the energies of the beta-particles emitted from a given nuclear species are not identical; it was first established by Chadwick in 1914 that the momenta of the β-particles, as measured by their curvature of path in a magnetic field, show a continuous distribution.

This continuous spectrum, upon which there may or may not

be superimposed sharp ' lines ' due to the subsequent process of ' internal conversion ', extends from zero (or as low as the apparatus can detect) to a definite upper limit of kinetic energy which may be well below 20 keV. (for H^3) or as high as 12 MeV. (e.g. for Li^8).

The shape of the distribution curve varies from one instance

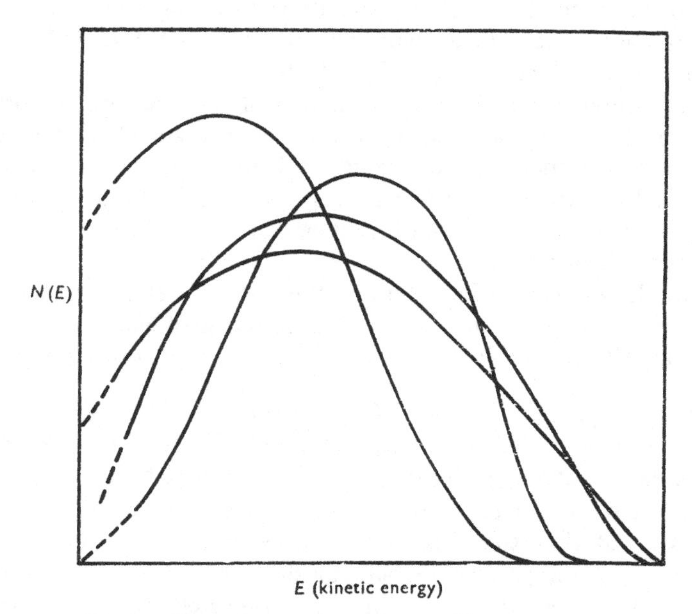

Fig. 9.—Typical approximate forms of beta-ray energy distributions

to another, but seems mainly to be determined by the limiting energy of the spectrum, the atomic number of the element concerned and (unless the atomic number is small) by whether positrons or negatrons are emitted.

These variations of shape, which are more evident if energy-distributions and not momentum-distributions are plotted, are illustrated in Fig. 9: curves obviously passing through or near the origin are characteristic of light nuclei, while heavy *negatron*-emitters give the more unsymmetrical energy-distributions with most particles having energies small compared with the upper limit. An extreme example (RaE) is shown in Fig. 15 (p. 56).

The continuous nature of β-ray spectra provided one of the most spectacular difficulties of nuclear physics.

Taking as an example † the successive transitions

$$\mathrm{AcU} \xrightarrow{\alpha} \mathrm{UY} \xrightarrow{\beta} \mathrm{Pa} \xrightarrow{\alpha} \mathrm{Ac}$$

we note that both AcU (U^{235}) and Pa yield a single discrete α-ray energy only, which is excellent evidence that all UY (Io^{231}) nuclei have the same energy content and all Pa nuclei have the same energy content; yet the transformation of the one into the other is accompanied by a β-ray having an energy anywhere between zero and an upper limit of about 0·2 MeV. To all appearances, either all but the slowest β-particles have acquired energy on the black market, or all but the fastest have lost energy in an equally irregular fashion.

The latter alternative seemed the more likely, and was in accordance with the fact that the difference of energy between ThC and ThD, as calculated from α-, β- and γ-energies in the two alternative paths $\mathrm{ThC} \xrightarrow{\beta} \mathrm{ThC}' \xrightarrow{\alpha} \mathrm{ThD}$ and $\mathrm{ThC} \xrightarrow{\alpha} \mathrm{ThC}'' \xrightarrow{\beta}$ ThD, can be consistently obtained only if the *upper limit* of the β-ray spectrum is regarded as the disintegration energy. If the principle of energy conservation is to be maintained, it must be assumed that the β-particle escapes from the atom with a variable part of the total disintegration energy: the remainder must appear elsewhere. Ellis and Wooster, by calorimetric measurements on RaE, showed that the missing energy was not retained by a thickness of 1·2 mm. of lead surrounding the source, nor was there any appreciable ionizing radiation escaping from the calorimeter; the heat given to the calorimeter agreed closely with that deduced from the *average* energy of the β-ray spectrum. Hence there is no possibility that the β-rays originally have identical energies and transfer a variable amount to atomic electrons, X-rays, neighbouring nuclei, or any carriers of energy that can be stopped by a reasonable thickness of material.

An equally important difficulty concerns the conservation of angular momentum: nuclei of even mass number all have integral spins in terms of the unit $h/2\pi$, while those of odd mass number

† Chosen for simplicity; even so, a slight complexity of the Pa spectrum is ignored.

have spins $(n + \frac{1}{2})(h/2\pi)$. In a β-ray change, the mass number is unaltered and hence the integral or half integral character of the spin does not change, yet the emitted electron has a spin of $\frac{1}{2}(h/2\pi)$.

Pauli suggested that the difficulties be formally resolved by assuming the fixed amount of disintegration energy to be shared in a continuously variable ratio between the β-particle and a *neutrino* which, while capable of carrying energy, momentum and angular momentum, has no charge and virtually no interaction with matter. The development of this hypothesis by Fermi and others will be considered later, but it should be pointed out immediately that on Pauli's hypothesis kinetic energy and momentum are shared between three bodies—the neutrino, the electron, and the nucleus: the conservation principles do not in such cases uniquely determine the energy and momentum of each particle (as they do in two-body problems), and different individual (identical) nuclei will distribute their equal available energies among the particles in a continuous range of different ways.

The recognition of the upper energy limit of the β-ray spectrum as the proper measure of the disintegration energy led Sargent (*PRS*, 139, 659) to attempt to correlate these upper limits with the corresponding decay constants for the natural β-radioactive bodies, in formal analogy with the well-known Geiger-Nuttall logarithmic plot of alpha-ray energies against α-ray decay constants. He found that the relation was less regular than for α-disintegration, but that almost all points fell near to one or other of two curves; the two lines do not (as in the α-ray case) correspond each to one radioactive family. As will be seen later, two or more curves are predicted by the theory, which distinguishes between 'allowed' and 'forbidden' transitions.

The relation between energy and lifetime depends quite strongly upon the atomic number of the nucleus concerned, but the natural β-ray bodies have all rather similar atomic numbers, with the exception of K^{40} and Rb^{87}, which in any case have half-periods (4×10^8 years and 6×10^{10} years) too far removed from the longest half-period among the heavy β-elements (RaD, 22 years) for the Sargent curves to be reasonably extrapolated to them.

The extension of the Sargent groupings to lighter elements is usually made in terms of the Fermi theory of β-disintegration, and will be discussed in due course (p. 44).

Simple and complex spectra

In many instances a β^--, β^+- or electron-capture process goes not to the ground state, but to an excited state of the product nucleus. If all the product nuclei are formed in the *same* state, a single definite amount of energy is shared between the β-particle and the neutrino: if, however, some transitions go to the ground state and some to an excited state, there will be two β-ray spectra of different limiting energies, having partial decay constants λ_1 and λ_2, so that the number of parent nuclei present decays exponentially with decay constant $(\lambda_1 + \lambda_2)$, and the two spectra decay together with a half-period $0{\cdot}693/(\lambda_1 + \lambda_2)$. It is clearly most important to be able to determine whether a β-ray spectrum, decaying with a single period, is simple or complex: that is, whether it corresponds to a single transition or to competing transitions from the same state of the parent nucleus to different states of the product nucleus. Three methods can be applied in various circumstances.

In the first place, the energy distribution may obviously depart from the smooth, single-peaked form characteristic of a single transition; such is the case when the two components are of roughly equal intensity, but considerably different energy.

In the second place, information may be obtained from the γ-rays and/or conversion electrons of the product element which accompany the β-ray change (or which, in the case of β-ray transition to a metastable state, are emitted later). If there are neither γ-rays nor conversion electrons, the β-ray transition presumably goes entirely to the ground state: if the γ-ray spectrum consists of a single line, and the number of γ-rays is equal to the number of β-rays, the transition goes entirely to an excited state: if, on the other hand, γ-rays or conversion electrons are found, but are together less numerous than the β-rays, the β-ray spectrum must be complex. Such a comparison of the numbers of β- and γ-rays usually involves the use of counters whose net efficiencies are determined by coincidence methods (p. 24), but if coincidence techniques are available a third and still better method of investigating the complexity of a β-ray spectrum can be used.

This is to investigate β-γ coincidences as a function of the thickness of an absorber placed between the source and the β-ray counter. These coincidences are due solely to the β-ray spectrum of *lower* energy, which leads to the excited state of the product nucleus: if therefore the number of β-γ coincidences per recorded γ-ray † is plotted against the thickness of the absorber, we shall obtain the absorption curve of the low-energy component of the β-ray spectrum. A refinement is to analyse the β-ray spectrum by a magnetic spectrometer and to observe β-γ coincidences for various parts of the β-ray spectrum. It is particularly valuable that the lower-energy component can be investigated in these ways, for it is obviously difficult to obtain a good value for its upper-energy limit by analysing the total β-ray spectrum into two 'smooth' components, even when some guide as to the shape of the separate components is available on theoretical grounds.

Energy considerations

On the neutrino hypothesis, the conversion of a nuclear neutron into a proton with the emission of a negatron and a neutrino is possible provided that

$$ {}_nM_Z^A - {}_nM_{Z+1}^A \geqslant (m + \mu) \quad . \quad . \quad . \quad . \quad (1), $$

where ${}_nM_Z^A$, ${}_nM_{Z+1}^A$ are the masses of the parent and product *nuclei*, m the mass of the electron and μ that of the neutrino.

The emission of a positron (and a neutrino) from the nucleus $(Z + 1, A)$ with reversion to the original nucleus Z, A is possible provided that

$$ {}_nM_{Z+1}^A - {}_nM_Z^A \geqslant (m + \mu) \quad . \quad . \quad . \quad . \quad (2), $$

since the mass of the positron is also m.

The masses quoted in tables are, however, not nuclear but *atomic* masses, thus for convenience the above inequalities should be rewritten in terms of the atomic masses. If we neglect the difference ‡ between the total binding energy of the electrons in (Z, A) and $(Z + 1, A)$, then $M_Z^A = {}_nM_Z^A + Zm$, $M_{Z+1}^A = {}_nM_{Z+1}^A + (Z + 1)m$, and the inequality (1) becomes

$$ M_Z^A - M_{Z+1}^A \geqslant \mu \quad . \quad . \quad . \quad . \quad . \quad . \quad (3). $$

† This proviso is necessary because the source is presumably decaying in time.
‡ At most a few kilovolts or say 1% of the rest-energy of the electron.

By a similar argument, we find that the condition for positron (plus neutrino) emission by the isobar $(Z + 1)$ is

$$M_{Z+1}^{A} - M_{Z}^{A} \geqslant (2m + \mu) \quad . \quad . \quad . \quad . \quad (4).$$

It will be noted that the asymmetry of equations (3) and (4) is due to the (conventional) use of *atomic* masses: the corresponding equations using nuclear masses, (1) and (2), are of the same form for both β^{+} and β^{-} transitions.

The capture of a K-electron by a nucleus is essentially an atomic process; the energy condition may be obtained by recognizing that the *atom* of mass M_{Z}^{A} emits firstly a neutrino of rest-energy μc^{2}, and then (in virtue of the filling of successive gaps in the electronic shells) one quantum each of K, L, . . . radiation of the product element. Since the successive X-rays would be replaced by a single one if an electron from outside the atom should drop straight in to the vacancy in the K-shell, the process will therefore be energetically possible if

$$M_{Z}^{A} - M_{Z-1}^{A} \geqslant \mu + E_{K}$$ [2]

where E_{K} is the binding energy of an electron in the K-shell of the product element, measured in atomic mass units.

It will be noted that immediately after the capture of the K-electron into the nucleus, the $Z - 1$ electrons remaining will be (crudely speaking) at incorrect distances from the new nucleus; there is thus a small uncertainty in the result of the present argument, comparable with the uncertainties already noted in respect of β^{+}- and β^{-}-emission.

Since E_{K} is only about $\frac{1}{5}mc^{2}$ even for the heaviest elements, K-electron capture is energetically more favourable than β^{+}-emission; moreover, the capture of an L, M, . . . electron is in principle possible, and will take over from K-capture, though with successively smaller transition probability, when the energy available is too small for the capture of a K-electron. In practice, therefore, the term E_{K} may be omitted from the last inequality.

Thus the conditions for interconversion by β^{+}-, β^{-}- and electron-capture processes between the two neighbouring isobars (Z, A) and $(Z + 1, A)$, are

$$Z \xrightarrow{\beta^{-}} Z + 1 : \qquad M_{Z} - M_{Z+1} \geqslant \mu \quad . \quad . \quad . \quad (5).$$

$$Z + 1 \xrightarrow{\beta^+} Z: \qquad M_{Z+1} - M_Z \geqslant (2m + \mu) \quad . \quad (6).$$

$$Z + 1 \xrightarrow{\text{capture}} Z: \quad M_{Z+1} - M_Z \geqslant \mu \quad . \quad . \quad . \quad (7).$$

Comparing the first and last of these conditions, we see that if the mass of the neutrino is negligible, two *neighbouring* isobars cannot both be stable, unless something other than energy requirements positively prevents a transition; both can be unstable, of course, witness the β^--chains which occur when

$$M_Z > M_{Z-1} > M_{Z-2}, \text{ etc.}$$

In fact, only three pairs of stable neighbouring isobars are known, namely Sb^{123}/Te^{123}, In^{113}/Cd^{113} and In^{115}/Sn^{115}. Very possibly, one of each pair may prove unstable, just as the discovery of β-activity in K^{40}, Lu^{176} and Re^{187} disposed of other examples. The neutron and the proton are, of course, neighbouring isobars, but free neutrons do not exist terrestrially and there is little doubt that the neutron is β^--active.

If we were to plot the atomic masses of all the isobars of one mass number against the atomic number Z we might obtain the result of Fig. 10 (*a*), where only one isobar is lighter than both of its neighbours and is therefore stable; on the other hand Fig. 10 (*b*) is also possible and explains the fairly frequent occurrence of stable isobars differing in mass number by two units. Fig. 10 (*b*) is typical of nuclei for which the mass number is even; the stable isobars are those for which the atomic number is even—that is, they contain even numbers of neutrons and of protons.

It is also obvious from Fig. 10 that at least one isobar of every mass number should be stable *against β-emission*. In fact, the only small or medium mass numbers missing in nature are 5 and 8; He^5 is presumably unstable with respect to neutron emission, Li^5 with respect to proton emission, while $Be^8 \rightarrow 2He^4$.

Since the average energy of β^-- or β^+-transitions, even for species near the bottom of curves such as Fig. 10, is of the order of 1 MeV. ($2mc^2$), it is unlikely that so few neighbouring isobars would be stable if the mass of the neutrino were appreciable compared with that of the electron.

In many instances (see *PPS*, 59, 408) the relative masses of the initial and final atoms are known sufficiently accurately, from nuclear reactions, for it to be shown that $M_Z - M_{Z+1}$ is closely equal to the maximum kinetic energy of the β^--spectrum; this is the most direct proof that the mass of the neutrino is small compared with that of the electron, but the proof is of limited accuracy since, as Kofoed-Hansen points out (*PR*, 71, 451), the end-point of the spectrum is in practice found by an extrapolation that assumes the neutrino mass to be zero.

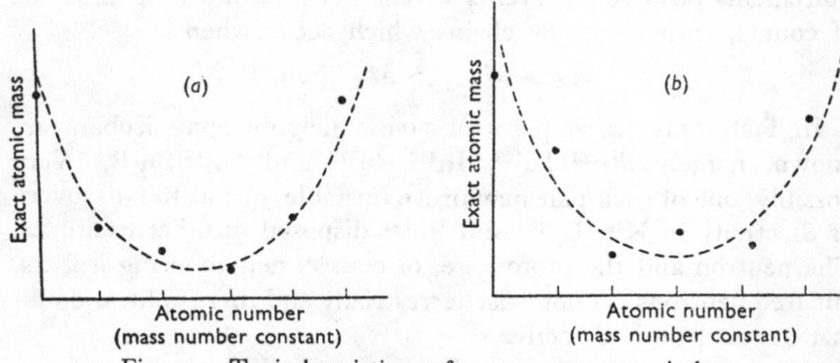

Fig. 10.—Typical variations of exact mass among isobars

Successive β-ray transformations

The very long chains of α- and β^--transformations which constitute the uranium-radium, thorium and actinium series cannot occur among lighter elements where α-radioactivity is impossible, but short chains of two successive β^--transitions are occasionally produced as the result of the addition of a neutron to the heaviest stable isotope of a light or medium element. As examples one may quote

$$\mathrm{Ca}^{48}_{20}\,(n,\gamma)\,\mathrm{Ca}^{49}_{20} \xrightarrow[2{\cdot}5h]{\beta^-} \mathrm{Sc}^{49}_{21} \xrightarrow[57m]{\beta^-} \mathrm{Ti}^{49}_{22}$$

$$\mathrm{Se}^{82}_{34}\,(n,\gamma)\,\mathrm{Se}^{83}_{34} \xrightarrow[30m]{\beta^-} \mathrm{Br}^{83} \xrightarrow[140m]{\beta^-} \mathrm{Kr}^{83}.$$

The abnormally high neutron-proton ratio of the first member of the chain is the essential requisite: an (n, γ) reaction is of course not the only way of increasing the neutron-proton ratio, and both Ca^{49} and Se^{83} may be produced by (d, p) reactions.

Pairs of successive β^+- and/or K-electron-capture transitions can similarly start from a nucleus of abnormally *low* neutron-proton ratio; for example, Ga^{65}, which may be produced from Zn^{64} by (d, n) or (p, γ) reactions, transforms successively to Zn^{65} and Cu^{65}:

$$Zn^{64}_{30}\,(d, n)\,Ga^{65}_{31} \xrightarrow[15m]{K} Zn^{65}_{30} \xrightarrow[250d]{K,\,\beta^+} Cu^{65}_{29}.$$

Another example is

$$Co^{55}_{27} \xrightarrow[18h]{\beta^+} Fe^{55}_{26} \xrightarrow[4y]{K} Mn^{55}_{25}.$$

The most striking β-ray chains among the lighter elements are those starting with the highly unstable nuclei produced in nuclear fission.

The neutron-proton ratio in the isotopes susceptible to fission is about 1·5, and although a few neutrons are emitted in the fission process, the fission fragments have a neutron-proton ratio of this same order, which is far too high for stability for nuclei of medium atomic number. The neutron-proton balance is set right by a series of β^--transformations, with the occasional intervention of neutron emission. Information about such chains is to be found in *RMP*, 18, 513: a typical pair of chains is †

$$U^{235}_{92}\,(n, \phi) \rightarrow Sr^{91}_{38} \xrightarrow[9\cdot7\,h.]{\beta^-} Y^{91*}_{39} \xrightarrow[51\,m.]{I.T.} Y^{91}_{39}$$

$$Sr^{91}_{38} \xrightarrow[9\cdot7\,h.]{\beta^-} Y^{91}_{39} \xrightarrow[57\,d.]{\beta^-} Zr^{91}_{40}\ \text{(stable)}$$

$$U^{235}_{92}\,(n, \phi) \rightarrow 2n^1_0$$

$$U^{235}_{92}\,(n, \phi) \rightarrow Xe^{143}_{54} \xrightarrow[1\,s.]{\beta^-} Cs^{143}_{55} \xrightarrow[\text{short}]{\beta^-} Ba^{143}_{56} \xrightarrow[0\cdot5\,m.]{\beta^-} La^{143}_{57} \xrightarrow[19\,m.]{\beta^-} Ce^{143}_{58} \xrightarrow[33\,h.]{\beta^-} Pr^{143}_{59} \xrightarrow[13\cdot8\,d.]{\beta^-} Nd^{143}_{60}\ \text{(stable)}$$

A pair of successive electron-capture transitions, namely $Ba^{131}_{56} \rightarrow Cs^{131}_{55} \rightarrow Xe^{131}_{54}$, has recently been reported (*PR*, 71, 382).

As an example of neutron emission we may quote

$$Br^{87}_{35} \xrightarrow[55\cdot6s]{\beta^-} Kr^{87}_{36} \xrightarrow[\text{instantaneous}]{n} \underset{\text{stable}}{Kr^{86}_{36}}.$$

† These chains are not given *as a pair* in the report quoted, and the pairing, though highly plausible, is a matter of hypothesis.

It should be noted that the emission of these ' delayed neutrons ' is not a ' radioactive ' process, but occurs because the product of a β-ray change (in this example, Kr^{87}) is formed in a state of high excitation from which the emission of a neutron is energetically possible, and forestalls the normal transition by γ-radiation to the ground state.

Owing to the relatively high stability (low mass) of nuclei containing even numbers of both neutrons and protons, and the relatively low stability (high mass) of nuclei having odd numbers of both particles, a nucleus containing an even number of protons and an odd number of neutrons has a stronger tendency to emit a neutron than its predecessor in the β-ray chain, which has an odd number of protons and an even number of neutrons, even though that predecessor has the higher neutron-proton ratio.

Theories of β-ray spectra

The detailed experimental results of β-ray spectroscopy are usually interpreted in terms of a theory which was, in its simplest form, built by Fermi (*ZP*, 88, 161) on the foundation of Pauli's neutrino hypothesis, but which is capable of many variations and can consequently be made to agree with a wide range of different experimental results. A full discussion was given by E. J. Konopinski (*RMP*, 15, 209; October, 1943), and no attempt will be made here to give more than the minimum account which will serve as a basis for discussion of experiments.

The transformation of a neutron to a proton (or *vice versa*), with the creation of a negatron (or a positron) and a neutrino was regarded by Fermi as in some ways analogous to the transition of an atom from one stationary state to another with the creation of a photon. The theoretical expression for the probability per unit time of the emission of a photon from an excited atom involves not only fundamental constants such as e^2, h and c, but also a factor characteristic of the initial and final states of the atom. The magnitude of this factor varies greatly according to whether or not the well-known selection rules for electronic transitions are obeyed; it thus expresses the great difference between ' allowed transitions ' and ' forbidden ' transitions, ' forbidden ' transitions being those which take place with much reduced but not zero probability.

Fermi's beta-transition equation contains analogous factors, but is naturally more complicated because not only do two particles share the energy of each transition, but one of them has an electric charge, and is therefore influenced by the Coulomb field of the nucleus from which it escapes.

For the particular case in which the mass of the neutrino is negligible in comparison with that of the electron, Fermi's equation is

$$N[W]dW \,.\, dt$$
$$= \frac{G^2}{2\pi^3} \,.\, |X|^2 \,.\, F[Z, W] \,.\, pW(W_0 - W)^2 \,.\, dW \,.\, dt \qquad (8),$$

where $N[W]dW \,.\, dt$ is the probability that in time dt a single nucleus shall emit a β-particle of *total* energy (kinetic plus rest energy) between W and $W + dW$, p is the corresponding momentum of the β-particle, and W_0 is the energy of transition—that is, the combined total energy of the β-particle and the neutrino. Since the neutrino is being assumed to have no mass and therefore no rest-energy, W_0 is the total energy of a β-particle at the upper limit of the spectrum. The units of W and W_0 are mc^2 ($= 0{\cdot}511$ MeV.), the unit of p is mc (corresponding to $B\rho = 1704$ gauss. cm.), and the unit of time is $h/2\pi mc^2 = 1{\cdot}29 \times 10^{-21}$ sec. The decay constant, λ, for the transition is $\int_1^{W_0} N[W]dW$, the half-period being, of course, $0{\cdot}693/\lambda$. Since $W^2 = 1 + p^2$, $WdW = p\,dp$ and an alternative form of equation (8) is

$$N[p]dp\,dt = \frac{G^2}{2\pi^3} \,.\, |X|^2 \,.\, F[Z, W]\,p^2(W_0 - W)^2 dp\,dt \quad . \quad (8a).$$

It will be convenient to consider the meaning of the various factors in reverse order.

The factor $pW(W_0 - W)^2$ represents the probability per unit energy range of sharing the total energy W_0 between the electron and the neutrino in such a way that the electron gets W and the neutrino $(W_0 - W)$. It can be derived from statistical arguments. It is not symmetrical with respect to W and $(W_0 - W)$, because the neutrino has no rest-mass, but it is zero when $W = 0$ or $W = W_0$ and has a maximum usually not far from $W = \frac{1}{2}W_0$.

The factor F represents the effect of the Coulomb field of the nucleus: positrons are repelled and negatrons are attracted, and consequently less very slow positrons and more very slow negatrons will be observed outside the atom than would be the case if the nucleus were uncharged; indeed, a positron which emerges with zero kinetic energy had negative kinetic energy when it was in the neighbourhood of the nucleus, and has had to escape, as do alpha-particles, through a potential barrier.

F, which is often called the 'Fermi function', is given by

$$F[Z, W] = \frac{(s+1) \cdot 2(2pR)^{2s-2} e^{\pi\alpha ZW/p} \,|\, \Gamma[s + i\alpha ZW/p] \,|^2}{(\Gamma[2s+1])^2} \quad (9),$$

where Z is the atomic number (strictly, that of the product nucleus) and is to be given a negative sign when positron emission is in question, R is the nuclear radius in units $\frac{h}{2\pi mc} = 3{\cdot}87 \times 10^{-11}$ cm. and $s = (1 - \alpha^2 Z^2)^{\frac{1}{2}}$, α being the 'fine structure' number $1/137$.

The factor $|\,X\,|^2$ is characteristic of the initial and final states of the nucleus; for allowed transitions it is of the order of unity, for forbidden transitions it has much smaller values and *may* be a function of W, in which event the spectral distribution will be different from that for an allowed spectrum with the same W_0 and Z. It plays a similar part in β-ray theory to that played by the 'multipole moment' in the theory of γ-ray transitions (see p. 77). The conception of a multipole moment, and the connection between the square of its magnitude and the rate of radiation of photons, are not very difficult to appreciate because we are familiar with the electromagnetic waves which a classical dipole radiates; we know, among other things, that they are vector waves, and we would not be surprised to find different formulæ for γ-ray transition probabilities if electromagnetic waves were scalar in nature.

The essential difficulty in β-ray theory is to guess the *kind* of interaction between the heavy particle (neutron or proton) and the light particles (β-particle and neutrino); it could be any one of five types, known as scalar (S), polar vector (V), axial vector (A), tensor (T), and pseudoscalar (P), or conceivably even a superposition of two types. The polar vector interaction was originally assumed by Fermi, by analogy with photon emission; Gamow and Teller have suggested, for reasons that will appear later, that the tensor interaction might be correct.

On Fermi's postulate, the transition is allowed (i.e., $|X|^2 \sim 1$) if both the angular momentum of the nucleus and its parity † are unchanged by the transition; on the Gamow-Teller hypothesis, an angular momentum change of one unit of $h/2\pi$ is also allowed. A physical interpretation of the angular momentum rules is that in the Fermi case the neutrino and the electron leave the nucleus with opposite spin, while in the Gamow-Teller case they may also leave with parallel spins. The much smaller value of $|X|^2$ when the angular momentum change exceeds the allowed amount corresponds to the relatively low probability that a particle shall leave a nucleus in such a way as to carry orbital angular momentum. Each unit of orbital angular momentum reduces the probability by a factor of the order of a hundred.

Provided the appropriate selection rules are obeyed, *every* interaction hypothesis gives $|X|^2 \sim 1$; that is to say, all forms of the theory predict about the same decay constant, and the same energy-distribution for an allowed spectrum with given values of Z and W_0; but the conditions under which the transition is allowed differ from one form of theory to another.

The factor $G^2/2\pi^3$, which we have left until last, expresses the 'strength' of the 'forces' responsible for β-decay. The factor is here written in the form used by Konopinski and, like all the other factors in equation (8), has been arranged to be a number. The corresponding equation of Fermi's original formulation employs a factor g^2 of the dimensions of (cm.^{3} erg)^2. These are not the only forms in which the fundamental constant of β-decay has been written; all forms are interconvertible by multiplying by factors involving π, h, m and c, appropriate compensation being of course made in other factors of equation (8).

The β-decay constant serves as a link between the β-emission forces and those responsible for the binding of neutrons and protons within a nucleus, in somewhat the same way that the electric charge e (or, if we prefer, the non-dimensional constant $2\pi e^2/hc = \frac{1}{137}$) relates the forces responsible for photon emission to the Coulomb forces that bind electrons within the atom. For the purposes of the present discussion, $G^2/2\pi^3$ is simply a constant the value of which, if required, must be obtained by comparing equation (8) with the results of experiment.

† See p. 78

Dependence of half-period upon transition energy; allowed and forbidden transitions

The first step in comparing theory and experiment is to see whether transitions are in fact classifiable into ' allowed ', ' first forbidden ', ' second forbidden ', and so on, the basis of classification being the relationship between transition energy and decay constant. Such a classification was made purely empirically by Sargent for the heavy natural β-ray bodies; in extending it to the whole periodic table, nuclei are first divided into three groups according to atomic number:

A. The lightest nuclei ($Z < 20$ or thereabouts) for which the nuclear charge has negligible influence on the escape of the negatron or positron and the Fermi function F may be taken as equal to one.

B. The nuclei of moderate Z (20 to 80), containing the majority of artificially-prepared β-active bodies; for group B the situation is rather complicated since the attraction or repulsion of the nucleus on the outgoing particle is neither negligible nor roughly constant for the diversity of elements concerned.

C. The heavy radioactive β-emitters ($Z > 80$). Here, although the influence of nuclear charge is very substantial, many β-ray bodies are known within a comparatively small range of Z, and the relations between their half-periods and transition energies may be compared statistically without much account being taken of the variation of Z within the group.

For the first group, the allowed and forbidden transitions can be distinguished at once by plotting the logarithm of the half-periods against the corresponding disintegration energies. Fig. 11 shows data for elements up to scandium and for values of W_0 between 2 and 13 (β-ray energies between about 0·5 and 6 MeV.). Where the decay of a nuclear species is ascribed to competing processes (β-rays to different levels of the product, β^+—β^- branching, etc.), each transition is plotted separately with the value of the half-period that would be found if it were the only process acting.

For example, Cl^{38} is now known (*AMAF*, 33A, No. 9) to decay by triple branching to the ground state and two excited states of A^{38}, the percentages of nuclei concerned being respec-

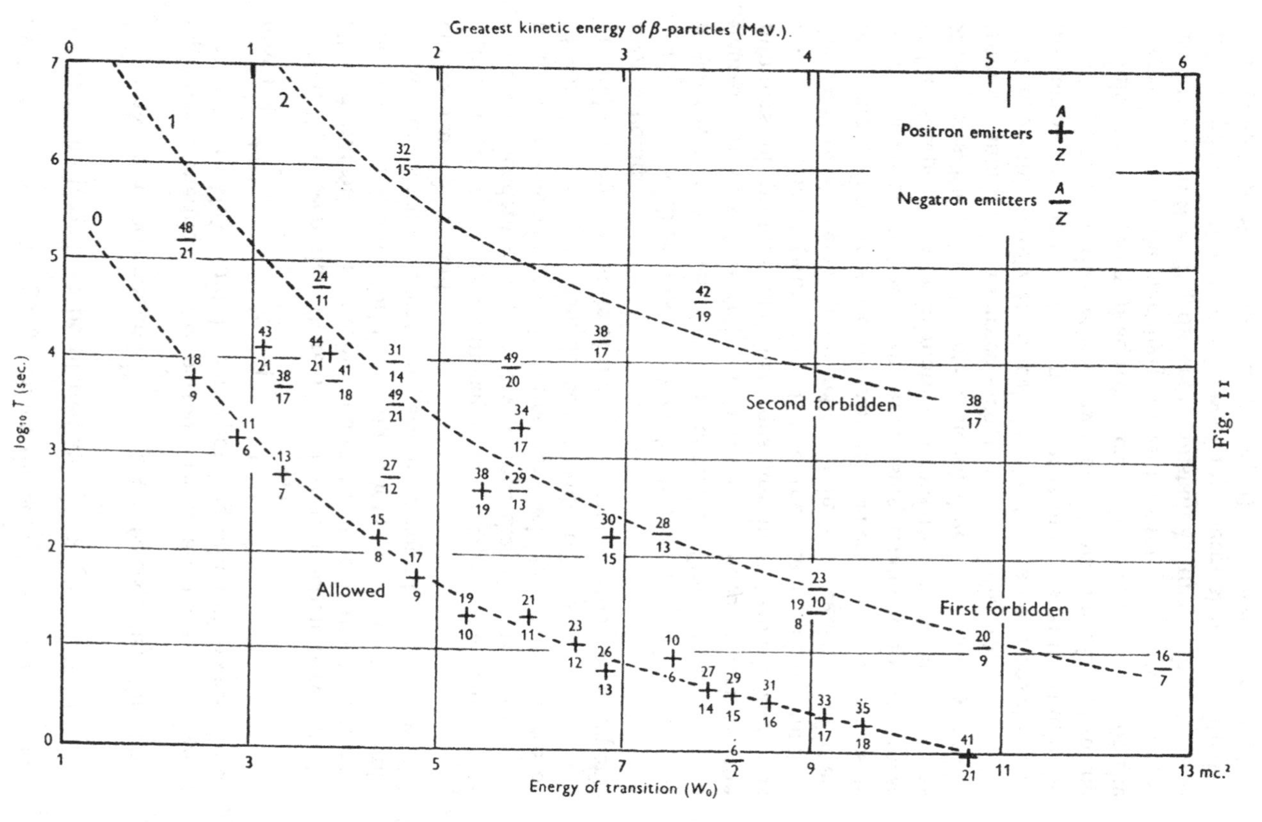
Greatest kinetic energy of β-particles (MeV.).
log₁₀ T (sec.)
Energy of transition (W₀)
13 mc.²
Positron emitters
Negatron emitters
Allowed
First forbidden
Second forbidden

Fig. 11

tively 53%, 11% and 36%. Since the actual half-period is 38·5 minutes, the periods to be plotted are $\frac{100}{53} \times 38{\cdot}5 \times 60$ seconds, $\frac{100}{11} \times 38{\cdot}5 \times 60$ seconds and $\frac{100}{36} \times 38{\cdot}5 \times 60$ seconds. The limiting kinetic energies of the partial spectra are respectively 4·94, 2·79 and 1·19 MeV., and the corresponding values of W_0 are therefore $1 + \frac{4{\cdot}94}{0{\cdot}511}$, $1 + \frac{2{\cdot}79}{0{\cdot}511}$ and $1 + \frac{1{\cdot}19}{0{\cdot}511}$.

It will be seen that there is a clear distinction between the allowed transitions, clustering round the line marked '0', and the first and second forbidden transitions, represented by only a few nuclei whose points are scattered round roughly parallel curves marked '1' and '2'.

Fig. 11 is, of course, equivalent to a Sargent diagram and is quite independent of any theory, except that the prediction of a factor of about 100 in half-period for each successive degree of 'forbiddenness' gives confidence to the identification of the various 'forbidden' lines in spite of the considerable scatter of the experimental points.

Fig. 11 also illustrates the remarkable regularity of increase of β^+-energy in the series of nuclei containing Z protons and $Z - 1$ neutrons from C_6^{11} to Sc_{21}^{41}. As a result of the β^+-transition, the numbers of protons and neutrons are interchanged, and this unique symmetry (which has led to the name 'mirror nuclei' for each pair) seems to be reflected in the regularity of their properties. The increase of transition energy as Z increases is of course associated with the increasing tendency for the most stable isotopes to contain more neutrons than protons. Thus as Z increases, the pairs of nuclei concerned get continuously farther from the minimum of curves such as those of Fig. 10.

Though this purely empirical classification is both possible and instructive for the very light nuclei, it is more usual to handle the problem in a semi-theoretical way that can also be adapted to the middle and the heavy β-active nuclei. The Fermi form of the theory is taken as the starting-point and, by integrating equation (8) from $W = 1$ to $W = W_0$, the decay constant λ—that is, the total probability per unit time that a β-ray of any energy shall be emitted—is obtained in the form $\lambda = K\phi[Z, W_0]$, where $\phi[Z, W_0] =$

$\int_1^{W_0} F[Z, W] \,.\, pW(W_0 - W)^2 dW$. The integral has to be evaluated by approximate methods, and values of ϕ are shown in Fig. 12 as functions of W_0 for (*a*) $Z = 0$, (*b*) $Z = 50$, (*c*) $Z = 85$.† Except for $Z = 0$, separate curves must be given for negatron and positron emission;‡ it will be remembered that to adapt Fermi's

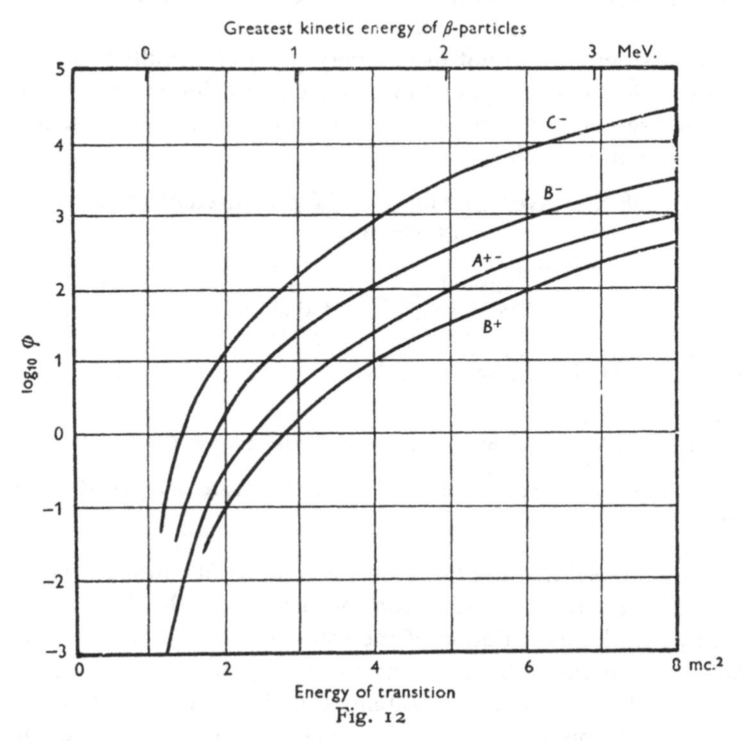

Fig. 12

formula for positron emission one replaces Z by $-Z$. The quantity ϕ is nearly proportional to the fifth power of the upper limit of β-ray energy, provided that this limit is considerably greater than $mc^2 = 0{\cdot}511$ MeV. (*PR*, 59, 63).

If the Fermi theory is correct, all allowed transitions should have the same value of ϕT, where T is the half-period and is equal to $0{\cdot}693/\lambda$; all first forbidden transitions should have ϕT

† These values, though based on information given in Konopinski's review (*RMP* 15, 209), are somewhat less accurate than his Table 2 and Fig. 3.

‡ No positron emitters of high Z are known; a curve for heavy positron emitters being thus of academic interest only, none is given in Fig. 12.

about 100 times greater, and so on. If one of the other forms of the theory (e.g. the Gamow-Teller tensor interaction) should be the correct one, the numerical values will be a little different, but the general behaviour will be much the same; since the theoretically undetermined constant G^2 is involved in each form of the theory, comparison of these predictions with observation will not tell us which form of the theory is right.

The predictions are not in fact very precisely fulfilled, but they are valuable for classifying the observed transitions; it is found that the great majority can be classified into the following scheme, where T is in seconds.

Table 2.—Average values of ϕT for different groups of β-transition

	Allowed Transitions (0)	1st forbidden (1)	2nd forbidden (2)
Very light nuclei (A) (Z < about 20)	Group 0A: 3×10^3	Group 1A: 2×10^5	Group 2A: 3×10^7
Medium nuclei (B) (Z 20 to 80)	Group 0B: 8×10^4	Group 1B: 2×10^6	Group 2B: 5×10^7
Heavy nuclei (C) (Z > 80)	Group 0C: 2×10^6	Group 1C: 2×10^7	Group 2C: 10^9

It will be seen that the Fermi formula does not completely represent the variation of transition probability with Z—if it did, there would be no variation of the value of ϕT as between allowed transitions in the light, medium and heavy groups A, B, C. On the other hand, it does give about the right variation of ϕ with energy, and about the right differences between the values of ϕ for negatron- and for positron-emitters.

Having in this manner identified allowed and forbidden transitions with considerable certainty among the light elements, if with somewhat less assurance for the medium elements, we can now enquire what selection rules are in fact obeyed by transitions which are identified as 'allowed'. The spins of radioactive nuclei are unfortunately seldom known, but a few instances seem definitely to violate the Fermi selection rule of no angular momentum change. For example, He^6, having even numbers of both neutrons and protons, presumably has zero angular momentum, yet it transforms at a rate which indicates an allowed transition to the ground state of Li^6 which is known to have spin 1; this transition is allowed by the Gamow-Teller rules.

Distribution of energy and momentum in allowed transitions

We have already seen that the energy-distribution of the β-rays varies considerably from one substance to another. According to the theory, it should depend, for allowed transitions at least, only upon the energy of transition (W_0), upon the atomic number of the product nucleus (Z), and upon whether positrons or negatrons are emitted. Comparison between theory and experiment is naturally laborious if the complicated theoretical prediction embodied in equations (8) and (9) has to be evaluated in full for each individual instance, and approximations are naturally welcome. Since some of them are instructive, they will be mentioned briefly before comparisons with experiment are considered in detail.

Near the upper limit of the spectrum, the factor $pW(W_0 - W)^2$ is the one which varies most rapidly with W, and so the number of particles per unit energy range will fall parabolically to zero as the end-point is reached, the tangent to the energy-distribution curve being horizontal at $W = W_0$. Had the mass of the neutrino been taken as finite, the theoretical curve would turn downwards as the end-point is approached and cut the energy-axis vertically. Although instrumental errors would to some extent blur such an effect, the most careful experiments in fact show no sign of it.†

At the extreme low-energy end of the spectrum, and for not too high values of Z, such that $p \ll Z\alpha$ and $(1 - \alpha^2 Z^2)^{\frac{1}{2}} \sim 1$, the Fermi factor F simplifies to K/p for negatrons and to $(K/p)e^{-2\pi\alpha ZW/p}$ for positrons, where K depends upon Z, but not upon p.‡ Now the 'statistical' factor $pW(W_0 - W)^2$ varies as p when p is small, W being practically constant at the value 1 since the kinetic energy of the β-particle is negligible in comparison with its rest-energy. It follows therefore that the number of *very* slow particles per unit energy range should be nearly independent of the energy when negatrons are in question, but

† It would have been observed in the spectrum of H^3 if the neutrino mass were as much as a three-hundredth of the electron mass (*N*, 162, 302).

‡ α is the fine-structure constant 1/137.

should vary as $e^{-2\pi\alpha ZW/p}$ for a positron spectrum. This would indicate that curves such as those sketched in Fig. 9 (p. 31) might be extrapolated to the origin for positron transitions, but that for negatron transitions they should cut the axis at a finite height. Experiments in the region of energy concerned (say, 50 keV. and below) are very difficult; some measurements extending into this region of energy, which do not agree with theory, will be mentioned on p. 62.

Approximations valid over the whole spectrum, except for this extreme low-energy range, may be made if Z is small. The obvious first approximation is to put $Z = 0$, when $F(W)$ becomes equal to 1 for all finite values of W; a better approximation, made by Kurie, Richardson and Paxton, gives the *momentum*-distribution in the form

$$N[p] = \text{constant} \times f[Z, p] \cdot (\sqrt{1 + p^2_{\max}} - \sqrt{1 + p^2})^2 \quad (10),$$

where $$f = p^2 \frac{x}{1 - e^{-x}} \text{ and } x = \frac{2\pi Z}{137} \cdot \frac{1 + p^2}{p}.$$

Thus a graph of $(N/f)^{\frac{1}{2}}$ against $W = \sqrt{1 + p^2}$ should give a straight line cutting the energy axis at $W_0 \equiv \sqrt{1 + p^2_{\max}}$. It must be noted that N is here the number of particles per unit range of momentum, and that a magnetic spectrograph, using a counter and a variable magnetic field, allows particles having a fixed *percentage* range of momentum to reach the counter; the rate of counting divided by the magnetic field is proportional to N.

Such a 'Kurie plot' is usually quite accurate enough when Z is below 30; when Z is high it is necessary to use the full Fermi function, plotting $\left(\frac{N[W]}{pWF}\right)^{\frac{1}{2}}$ or $\left(\frac{N[p]}{p^2F}\right)^{\frac{1}{2}}$ against W to get a straight line if the spectrum is of the 'allowed' form. The resulting diagram is often called a 'Fermi plot'. The labour of calculating the values for a Fermi plot can be greatly reduced by using numerical evaluations made by Bleuler and Zünti (*HPA*, 19, 375) and given by them in graphical form with a maximum error of 2%. Writing the energy-distribution function for negatrons as

$$N[W]dW = C[Z]W^2(W_0 - W)^2\, \theta[Z, W]dW$$

they give curves showing θ as a function of W for $Z = 40, 50, \ldots$

90, and $Z\theta$ as a function of W for $Z = 0$, 5, 10, 20, 30 and 40. For positrons, $\theta[Z, W]$ must be multiplied by $e^{-Z/P}$, where P is the value of $Z . \theta[W]$ for $Z = 0$. The square brackets indicate functions of the arguments within them; the reason for plotting $Z\theta$ when Z is small is that, as $Z \to 0$, $\theta \to \infty$, but $Z\theta$ remains finite and C varies as Z. To make a 'Bleuler-Zünti plot', one plots $\frac{1}{W}\left(\frac{N[W]}{\theta}\right)^{\frac{1}{2}}$ against W, remembering, as always, that the experiments give $pN[p]$ and not $N[W]$ directly.

Theoretical shapes of forbidden spectra

For forbidden transitions, the factor $|X|^2$ is not unity and may vary with W. The matrix element which we denote by X is sometimes composed of more than one term, each with its own influence upon the shape of the distribution curve and with its own selection rules; the relative magnitudes of these terms are quite unpredicted by the theory. In attempts to fit theory to experiment for forbidden spectra, a correction factor C is introduced which replaces the product $\frac{1+s}{2}|X|^2$. For the *first-forbidden* spectra, the following general rules are given by Konopinski.

He uses the symbol ΔJ for the change of total angular momentum of the nucleus; this is the notation of atomic spectroscopy, where the nucleus is treated as a single particle. In the next chapter we shall use L to denote this same quantity.

(1) T and A are the only interactions for which first-forbidden transitions can have $\Delta J = \pm 2$, and in this event the correction factor is of the form $(W_0 - W)^2 + W^2 - 1$. For high values of W_0, this modifies the allowed distribution by raising the ends relative to the middle. A very similar situation results from the T interaction with a transition in which $J = 0 \to J = 0$ and for the A interaction with $1 \longleftrightarrow 0$. For *small* Z, the S and P interactions give substantially the same behaviour, but for $Z \approx Z_c$ † the correcting factor increases monotonically with W.

(2) For heavy nuclei ($Z \gg Z_c$), the correction factor is independent of W for all possible interactions.

† In this connection, Z is regarded as "small" if it is distinctly less than $1{\cdot}6\ W_0$, which is denoted by Z_c. There seems to be no simple physical meaning to this criterion.

(3) With the V, T and A interactions, it is possible, though not necessary, for the correction factor to be independent of W even for light nuclei.

For *second-forbidden* transitions, and for still higher orders, different and about equally complicated conditions prevail, except that for second-forbidden transitions no single interaction can give an energy-independent correction factor with its consequent ' allowed ' spectral distribution.

Typical experimental results

In_{49}^{114}.

One of the most satisfactory tests of allowed β-ray transitions is that made by Lawson and Cork (*PR*, 57, 982) on In^{114}, the production and decay of which is illustrated in Fig. 13.

$$Cd^{113}(d, n)\,In^{114*} \xrightarrow[\text{50 d.}]{\text{0·19 MeV. conversion electron}} In^{114} \xrightarrow[\text{72 s.}]{\beta^-} Sn^{114}$$

Fig. 13

The high energy of the β-rays (end-point 1·98 MeV. kinetic energy, $W = 1 + \frac{1{\cdot}98}{0{\cdot}51} = 4{\cdot}88$) makes the spectrum easy to measure accurately, while the correspondingly short life (72 seconds) is no disadvantage since the β-ray source is continually replenished from the long-lived metastable state In^{114*}.

Fig. 14 (after Lawson and Cork's paper) shows that a very good straight line is obtained when $\left(\frac{N}{p^2F}\right)^{\frac{1}{2}}$ is plotted against the kinetic energy of the β-particles. The β-ray spectrum cannot be followed to very low energies because of the conversion electrons from the isomeric transition $In^{114*} \rightarrow In^{114}$. Note that F is the ' Fermi function '; the Kurie approximation is not sufficiently accurate for indium ($Z = 49$).

C_6^{11}.

This 20-minute β^+-emitter is readily prepared by bombarding boron with deuterons, according to the nuclear reaction $B^{10} + D^2 \rightarrow C^{11} + n^1$. It gives no γ-rays, so the transition $C^{11} \rightarrow B^{11}$ is simple, presumably going to the ground state. The values for

E_{max} of 0·981 MeV. (semicircular focusing method) and 0·993 ± ·01 MeV. (lens spectrograph) have been obtained by Townsend (*PRS*, 177, 357) and by Siegbahn (*AMAF*, 30A, No. 20) respectively. From these values, $W_0 = 2{\cdot}9$, and ϕ is about 3 (Fig. 12). Since $T = 1230$ sec., ϕT is about 3700 and the transition is an 'allowed' one—group 0A. Siegbahn, moreover, plots $\left(\frac{N(p)}{f}\right)^{\frac{1}{2}}$ against $B\rho$, f being the Kurie function (p. 50), which is quite accurate enough for so light a nucleus; a good straight

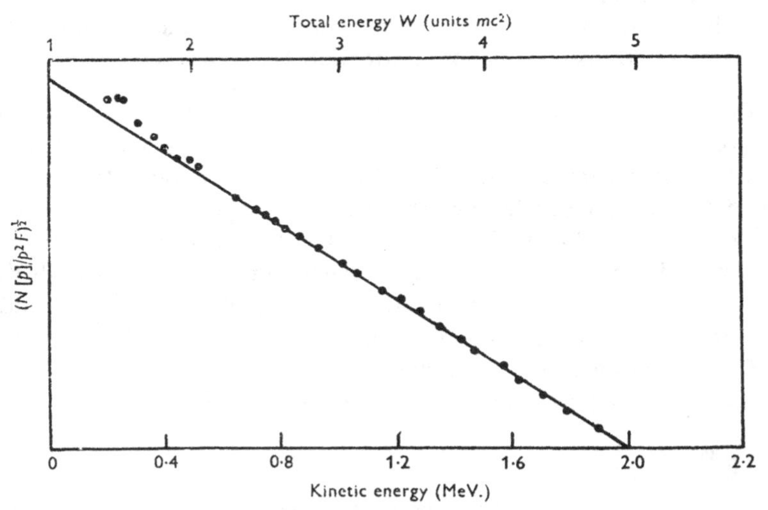

Fig. 14.—Fermi plot for In^{114}. (Lawson and Cork)

line is obtained for $W > 1{\cdot}4$ ($E > 0{\cdot}2$ MeV.). The excess of slow electrons can perhaps be explained as due to loss of energy by scattering in the source or in the source holder (an aluminium plate on which the boron was deposited).

Na^{24}_{11}.

Na^{24}_{11}, which emits negatrons and has a half-period of 14·8 hours, has been carefully studied by Itoh (*Proc. Math. Phys. Soc.*, Japan 23, 605), by Elliott, Deutsch and Roberts (*PR*, 63, 386), by Lawson (*PR*, 56, 131), and fairly recently by Siegbahn (l.c., and *PR*, 70, 127), following earlier work by Kurie, Richardson and Paxton. It seems clearly established that the upper energy limit is about 1·39 MeV. and that two γ-rays of energies 1·38 and 2·76 MeV. accompany the disintegration.

Coincidence measurements of the type described on p. 23 have shown (Feather and Dunworth, *PCPS*, 34, 442) that the spectrum is simple; Elliott, Deutsch and Roberts have found, by the method of p. 26, that one quantum of each of these γ-rays accompanies each β-disintegration: it is therefore to be presumed that all the β-transitions go to an excited state of Mg^{24}, with an energy of $1·38 + 2·76 = 4·14$ MeV. Later work (*PR*, 70, 985 and 72, 429) has confirmed this.

The spectrum being simple, the comparison of half-period with energy of transition is straightforward; it is found that the product ϕT is about 10^6, so that $Na^{24} \overset{\beta^-}{\rightarrow} Mg^{24}$ is a first-forbidden transition. In spite of this, evidence is accumulating that the spectrum is of the allowed form: Lawson found the Fermi plot to be substantially straight from the upper end-point down to 0·6 MeV., below which the ordinates lay above the line; Siegbahn in his first paper reports a straight line down to 0·4 MeV. and in his second paper, using an extremely thin source and source holder (total about 0·2 mg./cm.^{2}) to reduce the number of Compton electrons from the γ-rays (as also scattered β-rays), an excellent straight line down to less than 0·2 MeV.

K^{40}_{19}.

K^{40}_{19} is the naturally occurring β^--isotope of life $\sim 4 \times 10^8$ years, giving a spectrum which, as far as is known, is that of a single β-ray transition; though γ-radiation is observed, it is probably associated with an alternative transition to A^{40} by electron capture.

Dzelepow, Kopjora and Vorobjov (*PR*, 69, 538), analysing the very weak spectrum with a multiple 'semicircular' spectrometer, in which beams from six sources are focused on to a single counter, find the spectrum to be of a shape similar to an allowed transition, though they give no Fermi plot. This is an interesting transition, not only because it is particularly highly forbidden, but also because the nuclear spin of K^{40} is known to be equal to $4\frac{h}{2\pi}$. That of Ca^{40} is presumably zero, so the spin change in the transition is 4 units, which means a third or fourth forbidden transition on the Gamow-Teller selection rules and fourth or fifth on the Fermi rules. The upper limit is about 1·4 ($W_0 = 3·7$), and a comparison with the lifetime shows that the transition cannot be more than third forbidden: this is further evidence for the G.T. rules and against the Fermi rules.

P^{32}_{15}.

P^{32}_{15} is a β^--emitter of half-period 14 days, giving no γ-rays and having therefore a simple β-ray spectrum: the upper limit is 1·72 MeV., and the ϕT value identifies it as a second forbidden transition. Lawson (*PR*, 56, 131), using a sample weighing about 3 mg./cm.^{2}, found a straight Fermi plot down to about 0·8 MeV.: Siegbahn, with a thickness of the order of 0·2 mg./cm.^{2}, obtained a straight plot down to roughly 0·1 MeV.

It is interesting to note that none of the five possible interaction hypotheses predicts such an 'allowed' shape for a *second* forbidden transition, though it is possible that a suitable combination of two of them might be devised to do this.

RaE^{210} (Bi^{210}).

The spectrum of this 5-day β^--emitter has been studied many times with results that differ at low energies, but clearly indicate that the energy distribution is not of the 'allowed' form: the experimental difficulties are the usual ones of making sure that the source is thin enough and that scattering of the β-rays in the source holder and the material of the spectrograph is negligible. Fig. 15 gives the results of observations with semicircular-focusing magnetic spectrographs by Neary (*PRS*, 175, 71), by Flammersfeld (*ZP*, 112, 727), and by Alichanow, Alichanian and Dzelepow (*Phys. Zeit. Sowietunion*, 11, 204), and compares them with the Fermi 'allowed' distribution. The half-period is so long in comparison with the energy limit of the β-rays (1·17 MeV.) that the transition must be classed as first if not second forbidden. According to Konopinski (*RMP*, 15, 209) a suitable combination of vector and tensor interaction can explain both the lifetime and the energy distribution as measured by Alichanow *et al.*

We may sum up the situation by saying that most, though not all, β-ray spectra are roughly of the 'allowed' shape, but that no definite conclusion can be drawn as to which of the five interaction hypotheses is correct. For reasons given on pp. 48 and 54, the Gamow-Teller interaction is the present favourite.

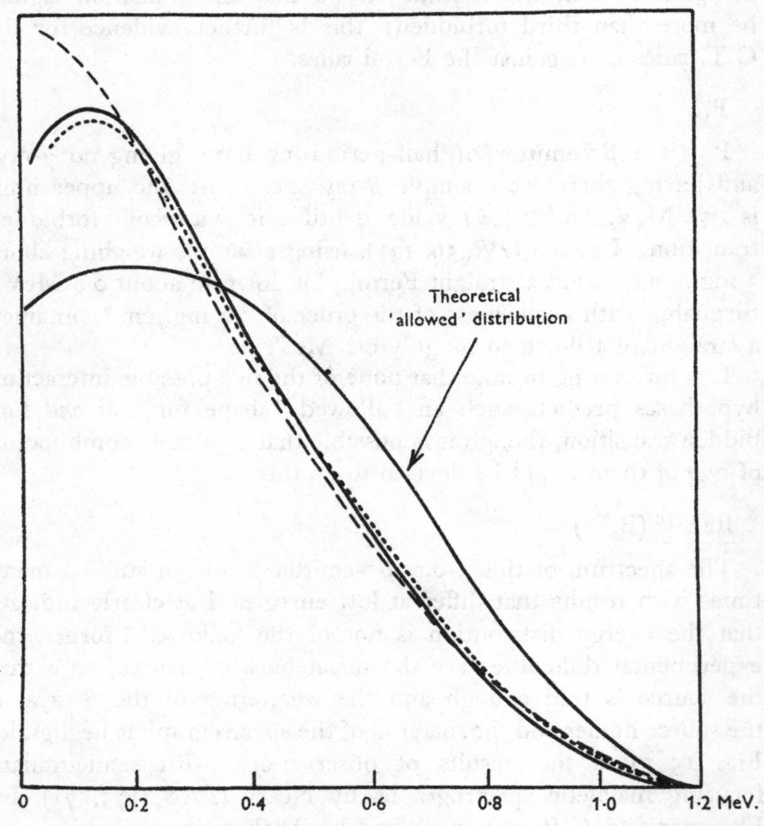

Fig. 15.—Experimental energy-distributions for RaE. Full line, Neary; broken line, Alichanian and Zavelsky; dotted line, Flammersfeld. Note the marked difference from the 'allowed' distribution of the Fermi theory.

Electron capture

We have already seen (p. 36) that the capture of an electron by a nucleus is an *energetically* more favourable process than the emission of a positron: we must therefore expect some nuclei to undergo spontaneous transitions which involve the emission of no detectable particle, but only the capture of an extra-nuclear electron and the emission of a neutrino.

The existence of this process was suggested by Yukawa and was observed by Alvarez (*PR*, 54, 486), who found that Ga^{67}_{31}, which can be prepared by bombardment of zinc with deuterons, protons or α-particles, decays with a half-life of 84 hours with the emission of the characteristic X-rays of zinc. The X-rays were detected with a counter, and were identified by the absorption-edge method: conversion electrons were also identified and indicate that the electron-capture transition proceeds to an excited state of the Zn^{67} nucleus. The nuclear process is therefore such as to *decrease* the nuclear charge by one unit: since no positrons are emitted, it must be a process of electron capture. There are actually some γ-rays and conversion electrons, and though the latter might at first be mistaken for negative beta-particles, that hypothesis would require the product nucleus to be Ge^{67}_{32} and is contradicted by the observation of *zinc* X-rays; it could also, of course, be disposed of by analysing the energy distribution of the electrons and finding that it is a line spectrum and not a continuous one and this was checked by Lyman.

Though energetically less favourable by $2mc^2$ or about 1 MeV., positron emission is intrinsically a more probable process since it involves an interaction between particles that are continuously present within the nucleus, whereas the capture of even a K electron depends upon the small but finite probability that this electron shall be 'within' the nucleus. Hence as soon as the energy available for the K-capture process substantially exceeds $2mc^2$, positron emission tends to be dominant. This dominance is sometimes offset by the β^+-transition to the ground state being forbidden, and the energy available being insufficient to permit a β^+-transition to a higher state that is 'allowed'. The K-capture transition is governed by the same selection rules, since a K electron has no orbital angular momentum, and indeed if it had orbital angular momentum would have a wave function of zero amplitude at the nucleus. With its extra 1 MeV. available energy the K-capture transition has a much wider choice of levels to which to go, and will almost certainly be able to make an 'allowed' transition to one of them. There are thus many cases where positron emission, though energetically quite possible, is beaten in competition by K-capture to an excited state, followed by the emission of gamma-radiation.

Theory of the electron-capture process

The theoretical study of the K-capture process has followed similar lines to those followed in the theory of beta-emission: indeed positron emission can be considered formally as the capture of an electron from the continuum of negative-energy electron states postulated by Dirac, the positron being the vacancy left in the continuum. The final formulæ for the transition probability are however simpler in the case of electron capture, since in this process the momentum is shared between *two* particles only (the neutrino and the nucleus) and there is no complication of the sharing of the corresponding energy in a continuously variable manner with a third particle. For allowed transitions the result is

$$\lambda = \frac{G^2}{2\pi^3} \mid X \mid^2 . \, 2(\alpha Z)^{2s+1}(2R)^{2s-2}\left(1 + \frac{s}{\underline{|2s}}\right) W^2,$$

where the symbols have the same meanings as in the formulæ given earlier for β-decay, except that W is the constant energy of the transition and not the variable energy of the β-particle.† This energy cannot be determined by direct experiment since (apart from the negligible recoil energy of the atom) it is carried away by the unobservable neutrino. It can, however, be inferred from the masses of the parent and product atoms, if these are known accurately; or if some of the nuclei concerned decay by positron emission it can be inferred from the energy limit of the positron spectrum. If W_{β^+} is this upper energy limit (in units of mc^2, as usual, and including the mass energy of the positron), then $W = W_{\beta^+} + h\nu_{\beta^+} - h\nu_K + \eta$, where $h\nu_{\beta^+}$ and $h\nu_K$ are the energies of such γ-rays as may follow the alternative β^+- and K-transitions and η is the mass energy of the electron minus its binding energy in the atomic shell. For capture from the K-shell (which alone has been experimentally observed, though capture from other shells is possible with the lower probability appropriate to the smaller value of the electron wave function at the nucleus), η is equal to the quantity $s = (1 - \alpha^2 Z^2)^{\frac{1}{2}}$ which occurs elsewhere in the theory of these processes. An approximate formula is

† Since s is not integral the factorial must be interpreted in terms of a gamma function.

$\lambda = \frac{G^2}{2\pi^3} | X |^2 . 2(\alpha Z)^3 W^2$; this is valid if s is near to unity: that is, if the binding energy is small compared with the mass energy of an electron (0·51 MeV.). This condition is tolerably well satisfied even for quite heavy nuclei.

For an Lth forbidden transition, a method due to Bethe (*RMP*, 8, 82) gives

$$\lambda = \frac{G^2}{2\pi^3} | X |^2 . \frac{2\pi}{9} (\alpha Z)^5 (W_{\beta^+} R)^{2L-2} \frac{(W_{\beta^+} + L)^2}{1^2 . 3^2 \dots (2L - 1)^2 . L} \quad (11).$$

It will be noted that the same constant G and matrix elements $| X |^2$ enter into these expressions as into those for β^+- or β^--decay: the ratio of K-capture to β^+-emission therefore provides a simple if not particularly stringent test of the theory. This ratio, as Good, Peaslee and Deutsch point out, can be determined experimentally with considerable certainty in favourable cases (*PR*, 69, 313).

When, as for example with V^{48} or the 6·5-day Mn^{52}, the alternative positron emission and K-capture processes occur in comparable numbers, some care may however be required in establishing the existence of the electron-capture process. Thus, if the β^+-process goes to an excited state of the product nucleus and the subsequent transition to the ground state goes by way of internal conversion, the ejection of the conversion electron leaves a vacancy in the K-shell and leads to the emission of X-rays identical with those which follow the capture of a K-electron by the original nucleus.

As an example of the experimental methods of elucidating such competitive transitions, we may summarize the experiments made with Mn^{52} by Peacock and Deutsch (*PR*, 69, 306) and by Good (*PR*, 69, 313). This is a 6·5-day positron-emitter.

The positron spectrum was first investigated and found to be of the 'allowed' form with an energy limit of 0·58 MeV.,† also the number of β^+-γ coincidences per recorded positron was $4{\cdot}2 \times 10^{-3}$ and did not change when an absorber was placed between the source and the β-ray counter. Thus it was clear that the β^+-transitions go to a single level of the resulting Cr^{52} nucleus. The photoelectron groups ejected by the Cr^{52} γ-rays

† This value, obtained with a β-ray spectroscope, is presumably more accurate than earlier cloud-chamber measurements (p. 12).

from a radiator of lead indicated γ-ray energies of $0{\cdot}73_4$, $0{\cdot}94_0$ and $1{\cdot}4_6$ MeV., together with annihilation radiation. The known efficiencies of the γ-ray counter for γ-rays of these three energies were $1{\cdot}10 \times 10^{-3}$, $1{\cdot}25 \times 10^{-3}$ and $1{\cdot}80 \times 10^{-3}$, and their sum is equal to the relative β-γ coincidence rate observed, thus it was evident that each of these quanta is emitted after every β-dis-

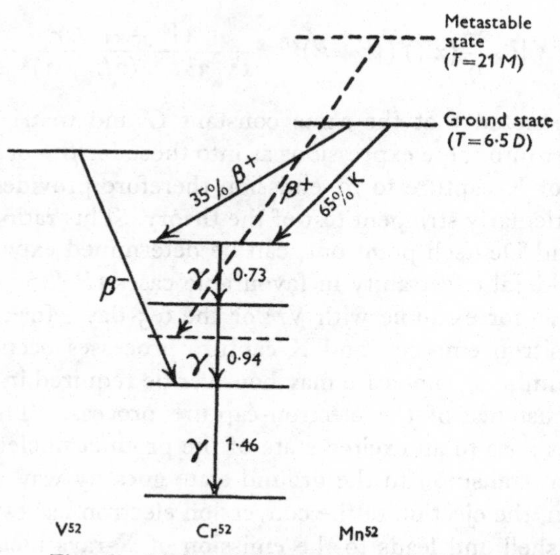

Fig. 16.—Mn^{52} – Cr^{52} transitions. The broken lines indicate the transition from the metastable state (p. 71); later work (*PR*, 71, 467) has shown that it goes to the 1·46 MeV. level of Cr^{52}

integration. X-rays following the alternative electron-capture process were observed, and the number of X-γ coincidences per recorded X-ray was also $4{\cdot}2 \times 10^{-3}$, indicating that the electron-capture transition goes to the same energy level of Cr^{52} as does the β^+-transition. The ratio of β^+-transitions to electron-capture transitions was determined to be 0·35/0·65, by comparing the number of annihilation quanta with the number of γ-ray quanta. For details of this comparison, see *PR*, 59, 313.

The decay scheme is as shown in Fig. 16. It may be taken as well established except that the order of the γ-rays is uncertain, though the 1·46 MeV. γ-ray is presumably the lowest one since a single γ-ray of about that energy is known to follow the transition $V^{52} \xrightarrow{\beta^-} Cr^{52}$. For the decay of an isomer of Mn^{52}, see p. 71.

The β^+-energy and the half-life indicate that the transition is either allowed or first forbidden. The fact that the spectrum has the allowed form does not necessarily mean that it is an allowed transition, for the tensor, vector and axial vector interactions can, by suitable choice of matrix elements, be made to predict substantially the 'allowed' form for a first-forbidden transition. As stated above, the experimental percentage of positron emission is 35%. The tensor interaction, which at present is the popular favourite, predicts 36%, 32% and 37·5% positron emission according to whether the transition is allowed, first forbidden with zero angular momentum charge, or first forbidden with one unit change of angular momentum. But if the transition is allowed, any of the interactions predicts a percentage in agreement with that observed experimentally. Thus, as usual, the general theory is supported, but little discrimination can be made between the various interaction hypotheses.

Competition between β^+-, β^--, and K-capture processes

It was first pointed out by Sizoo that if two isobars $(A, Z - 1)$ and $(A, Z + 1)$ are both stable, it is almost certain that the intervening isobar (A, Z) is energetically capable of decaying to both of these stable nuclei by electron capture and by β^--emission respectively; this would be quite certain if the neutrino mass were known to be accurately zero and if the binding energy of the electron were negligible. The reasons for the above statements will be evident from the discussion on p. 35, from which it will also be clear that if the *nucleus* (A, Z) has a mass which exceeds that of each of its neighbours by more than the mass of an electron, triple branching between β^+-, β^-- and K-capture is to be expected: it will, of course, be exceptional for the three transition probabilities to be so nearly equal that each of the three processes is experimentally observable.

The best-known example is the 12·8-hour Cu^{64}, which may readily be made from copper by $(n\gamma)$ or (dp) reactions, the 5-minute Cu^{62} being allowed to decay before observations are started. Cu^{64} emits positrons of limiting energy 0·66 MeV., negatrons of

0·58 MeV., and X-rays characteristic of nickel. Since no γ-rays or conversion electrons are observed, the only explanation for these X-rays is that some Cu^{64} nuclei transform to Ni^{64} by K-electron capture. Negatrons, positrons and X-rays are as 2 : 1 : 2, so the half-periods of decay by β^--emission alone and by β^+-emission alone would be about 32 and 64 hours respectively,† and multiplying these times (in seconds) by the appropriate values of ϕ, for a group B nucleus, as obtained from Fig. 12, we find values of ϕT equal to about $2{\cdot}2 \times 10^5$ and 7×10^4, so from Table 2 both transitions fall into the allowed group 0B.

The β^+- and β^--spectra have been investigated by various workers, for example by Backus (*PR*, 68, 59), whose measurements relate specifically to the low energy end of the spectra, and extend from about 50 keV. down to about 10 keV. For such low energies, a spectrograph using electrostatic deflection is practicable, and Backus used the 127° focusing method of Hughes and Rojansky (*PR*, 34, 284). The source, prepared by the Zn^{66} (d, α) Cu^{64} reaction, was deposited on a 0·03μ collodion film by the evaporation of a drop of copper nitrate solution, and was placed obliquely in front of the entry slit of the spectroscope. Beyond the exit slit was a Geiger-Müller counter separated from the spectroscope by a 0·08μ collodion window very thinly silvered and supported on a metal grid; this window could withstand the difference of pressure (about 5 cm. of mercury) between the counter and the vacuum. To reduce scattering from the deflecting electrodes, these were in the form of grids instead of the usual solid electrodes.

The number of negatrons per unit range of energy was found to be nearly proportional to the kinetic energy, in sharp contrast to the theoretical prediction (p. 49) of a nearly uniform distribution of slow negatrons on an energy plot. For the positrons, on the other hand, the agreement with theory was quite good; it will be remembered that for positrons the low energy end of the spectrum should have the form $Ke^{-2\pi\alpha ZW/p}$. It has been pointed out by Lewis and Bohn (*PR*, 69, 129) that no single one of the five interaction hypotheses can explain these observations, though they could be explained by suitable combinations of two inter-

† An infrequent γ-ray later observed (*HPA*, 19, 219) does not affect the argument.

actions—a device which is theoretically permissible, but distastefully artificial.

Cook and Langer (*PR*, 73, 601) have since investigated the Cu^{64} spectra with a large magnetic spectograph, effects of scattering in the source and by slits being kept very small by careful design and construction. They find that, both for electrons and positrons, there is an excess of low-energy particles as compared with the Fermi distribution law, which is well obeyed above 0·27 MeV. for the positrons and above 0·19 MeV. for the negatrons.

The same authors (*PR*, 74, 227) have investigated the positron spectrum of Cu^{61} (allowed transition, upper limit 1·2 MeV.) with the same instrument, and find an excess of lower-energy particles extending up to about 0·5 MeV. The energy below which such deviations occur is thus not constant, even for nuclei of the same element, and is therefore unlikely to be due to any unforeseen effect of the Coulomb field of the nucleus. The deviations seem rather to occur at energies below about 0·4 of the upper limits of the spectra. Thinner sources seem to reduce them (*PR*, 74, 847).

From what has been said in the present chapter about experimental and theoretical results concerning β-ray spectra, the following summary of the situation may be set down.

(1) The neutrino hypothesis provides the only known means of accounting for the general nature of β-ray spectra; it is further supported by the experimental facts of electron-capture, as far as these are known.

(2) The grouping of β-ray transitions into allowed, first forbidden, second forbidden, etc., is a well-established fact, even though a number of ' borderline cases ' exist.

(3) An interpretation of this grouping by selection rules can be made with good if not perfect consistency if the rules are those proposed by Gamow and Teller; the Fermi selection rules do not fit the facts so well.

(4) Except at comparatively low energies, where spectra (and experimenters) differ, the observed energy distribution, even for ' forbidden ' transitions, is close to that predicted by any of the detailed forms of the theory for *allowed* transitions. The more careful the experiments, the more nearly true this appears to be. RaE, however, has definitely not the ' allowed ' distribution.

(5) Attempts have been made to explain the observed decay

periods and detailed energy distributions by combinations of the five possible interaction types. With this amount of latitude, almost any observed spectrum can be fitted, thus isolated successes have no significance; there is no evidence that any of these attempts can consistently explain such deviations from the allowed form as may exist, and the interest of such attempts diminishes with each new careful investigation of a ' forbidden ' transition, in which an ' allowed ' shape is observed.

Difficult as the experiments are, there is distinct evidence that even ' allowed ' spectra are not of the Fermi form at low energies; it is, however, difficult to feel quite confident that any of the theories, even if its basic assumptions regarding interactions happen to be correct, can be expected to deal adequately with the extremely low-velocity end of the spectrum where the β-particle energies are comparable with the binding energies of the atomic electrons.

The situation is, at least in principle, still further complicated by the introduction of the meson theory of nuclear processes, in which forces between nucleons are associated with the interchange of mesons (positive, negative or neutral) between the nucleons, and β-emission is regarded as the breaking up of one of these mesons into a neutrino and a β-particle. Since the meson theory affects only the multiplying constant G^2 in equation (8) (p. 41) there is no need to attempt any discussion here, except to remind the reader that the radioactive decay of free mesons (mean life about $2{\cdot}2 \times 10^{-6}$ sec.) † is clearly established by experiments in the field of cosmic radiation (e.g. *PR*, 64, 199; *PR*, 70, 859).

It is thus not possible at the present time to give a comprehensive account of β-ray processes: the experiments are not yet sufficiently extensive and reliable, while the theories are capable of almost infinite variation.

Recoil of nuclei on β-emission or electron-capture

Although the problems of β-ray spectra seem quite insoluble without the basic assumption of neutrino emission, it is clearly important to seek for more direct evidence of the existence of

† Only about half of the cosmic ray mesons brought to rest in medium and heavy materials give decay electrons; these are apparently the positive mesons, the negative ones being attracted into and captured by nuclei before they have time to decay.

neutrinos. The most obvious line of attack is to measure the energy distribution of the nuclear recoils that accompany β-ray emission, for whatever momentum is carried off by the neutrino must be reflected in recoil momentum of the nucleus and must be added vectorially to the recoil momentum that reflects the momentum of the β-particle. Thus if the energies of the β-particle and the recoil nucleus are observed (for example with a radioactive gas in a cloud chamber), the corresponding momenta should not be equal, except when the β-particle happens to take the limiting spectral energy and the neutrino takes none. Unfortunately the energy of recoil is small, even for light nuclei. The most favourable β-radioactive gas is He^6, where an upper limit of recoil energy about 1600 eV. is to be expected; but He^6 has a half-period of about 0·8 second, and it would clearly be very difficult to prepare and extract this substance (e.g. by bombarding beryllium with fast neutrons) and introduce it into a cloud chamber before it had practically all decayed.

Experiments on recoils in a β-active gas were made with Cl^{38} by Crane and Halpern (*PR*, 53, 789; 56, 232); Cl^{38} has a convenient period (37 minutes) and a high limiting β-particle energy (5 MeV.), but unfortunately the β-ray spectrum is complex, gamma-rays of 2·25 and 1·65 MeV. being emitted following the less energetic β-ray transitions of 2·75 and 1·1 MeV. energy. The most favourable disintegrations for observing neutrino momentum would be those of the '5 MeV.' group in which the β-particle energy is low and nearly all the energy is taken by the neutrino, but then one cannot be sure that these are not transitions of 1·1 MeV. total kinetic energy, in which event γ-ray recoil will intervene. In any event the recoil is too small to be observed as a track, but only as a bunch of droplets at the beginning of the β-ray track. By choosing disintegrations which had occurred at a suitable time before the expansion, so that the originally very closely spaced ions had diffused apart somewhat before being 'fixed' by the condensation, Crane and Halpern were able to count the droplets and to use their number as a measure of the energy of recoil; the number of 'recoil droplets' at the beginning of a β-ray track of the full 5 MeV. energy was used as a standard of reference, since in that case the recoil is of known momentum, equal to that of a 5 MeV. β-particle. The momenta of the individual beta-particles were measured by the curvature of the tracks in a magnetic field.

The results were inconsistent with the recoil being due to β-particle momentum only and so support the neutrino hypothesis, though the experimental difficulties were such as to prohibit a thorough comparison with the consequences of that hypothesis.

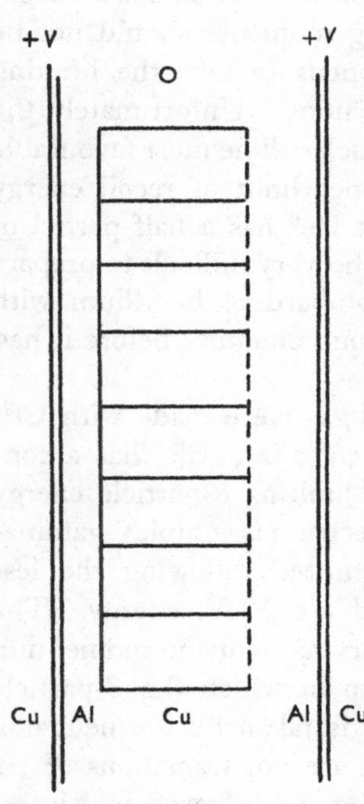

Fig. 17.—Jacobsen and Kofoed-Hansen's recoil experiment

Another experiment on β-ray recoil has since been made by Jacobsen and Kofoed-Hansen (*KDVS*, 23, No. 12),† who studied the recoil of Rb^{88} on its production by the β-decay of Kr^{88}. Since Rb^{88} is itself β-active, the place where a recoiling Rb^{88} ion comes finally to rest can be identified by the subsequent emission of the second β-particle from that place. In these experiments, a rectangular copper box with internal partitions and one gauze side was placed midway between two aluminium foils (Fig. 17) backed by copper plates, the whole being immersed in gas at low pressure containing Kr^{88}; a potential difference, variable from one experiment to another, was maintained between the box and the surrounding electrodes. After a suitable exposure of the aluminium foils to such Rb^{88} recoil ions as were able to reach them against the opposing electric field, they were removed from the apparatus and their Rb^{88} β-activities compared, by Geiger counters, with that of a small negatively-charged brass disk which was used as a monitor of the total amount of exposure to Kr^{88} disintegrations. If A_1, A_2 and A_3 are respectively the activities of the right- and left-hand foils and the brass disk, then $A_1 - A_2$ is proportional to the number of recoil ions that emerge

† A further account of this work is given in *PR*, 73, 675.

from the inside of the box and reach the right-hand foil; since A_3 is proportional to the exposure, $\frac{A_1 - A_2}{A_3}$ is proportional to the chance that an average recoil ion shall pass through the opposing potential difference V. Since Rb, as an alkali metal, has a very low first ionization potential, but a high second ionization potential, it was not expected that doubly-charged ions would be formed and so complicate the results.

When $\frac{A_1 - A_2}{A_3}$ was plotted against V for a series of separate measurements with different values of V, a curve of the form of Fig. 18 was obtained, showing that no recoil ions can overcome a potential difference exceeding about 51 volts. Now the limiting β-ray energy from Kr^{88} is 2·43 MeV., and calculation shows that the energy of recoil should then be 51·2 eV.; the end-point of the retarding-potential graph is thus in excellent agreement with the hypothesis that when the β-particle has the greatest possible momentum, the neutrino has none, and confirms the expectation that all the Rb^{88} is in the form of $(Rb^{88})^+$ ions. The actual form of the graph is less easy to interpret, for two main reasons. Firstly, the graph of Fig. 18 is not a true picture of the energy distribution of the recoil ions, owing to the obliquity of the directions of recoil of some ions with respect to the direction of the electric field. The true integral energy-distribution curve must lie somewhat above the experimental curve at all points between $V = 0$ and $V = 51$ volts. It is to remove many obliquely-travelling ions that the partitions are placed within the box. Secondly, even if the true recoil energy distribution were known, it would not be possible to deduce uniquely the momentum distribution of the β-particles and neutrinos. All that can be done is to make various hypotheses about the β-ray process and see whether they are consistent with the observations. Jacobsen and Kofoed-Hansen find:

(1) That the recoil energy distribution calculated from the known momentum distribution of the β-particles alone lies wholly *beneath* the experimental curve. Even if allowance is made for γ-ray recoil (the β-spectrum is complex) the discrepancy remains, and the experiments therefore strongly indicate that momentum leaves the nucleus in some other form.

(2) If Pauli's neutrino hypothesis be adopted, the recoil dis-

tribution can be explained only if there is a tendency for β-particle and neutrino to emerge in the same general direction.

The experiment is not easy to perform in a reliable way and the authors are properly cautious about the second conclusion, but it is clearly a most interesting result. Crane and Halpern's work also showed rather more neutrinos in the 'forward' than in the 'backward' hemisphere. Recoils from P^{32} and Y^{90} (*PR*, 73, 216 and 1173) seem to show most neutrinos in the back-

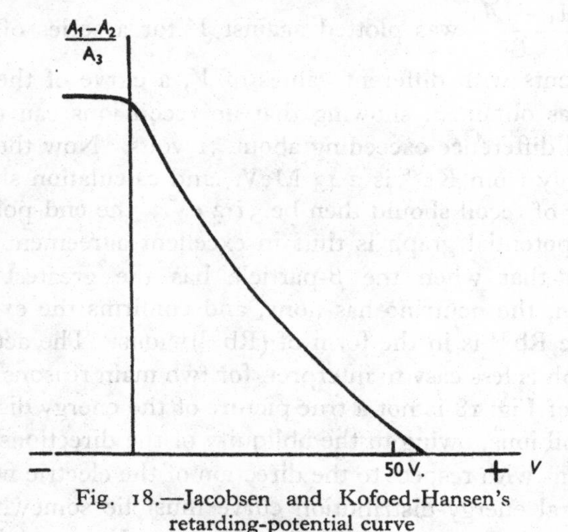

Fig. 18.—Jacobsen and Kofoed-Hansen's retarding-potential curve

ward hemisphere. Only forbidden β-transitions combining high energy with conveniently long life have so far been studied; for these, the theoretical predictions are complicated (*PR*, 71, 456).

The recoil of Li^7 on its formation by K-capture from Be^7 has been studied by Allen (*PR*, 61, 692) in what is the most satisfactory experimental test of the neutrino momentum. In such a pure K-capture process no recoil at all would be observed if it were not for the neutrino (the X-ray recoil being quite negligible). The mass of the Be^7 atom, as deduced from nuclear disintegration data, is 7·01916, while that of the stable Li^7 is 7·01822, and the energy available for the transition is therefore $0{\cdot}00094Mc^2$, where Mc^2 is the energy corresponding to the destruction of one atomic mass unit. If the transition goes to the ground state of Li^7, and if the rest-mass of the neutrino is zero, the neutrino will take practically all this kinetic energy. The momentum of the neutrino is thus $0{\cdot}00094Mc$: this is equal to the recoil momentum of the Li^7, which having mass $7M$ therefore has recoil energy

$$\frac{1}{14M}(0{\cdot}00094Mc)^2 = \frac{(\cdot 00094)^2}{14}\frac{M}{m}\,.\,mc^2 = \frac{8{\cdot}80 \times 10^{-7}}{14} \times$$

$1840mc^2 = 1{\cdot}17 \times 10^{-4}\,mc^2 =$ about 60 eV.

The problem of detecting, and measuring the energy of, such slow recoil particles was ingeniously solved by Allen. He extracted an apparently large fraction of the radioactive Be^7 from the LiF (in which it was produced by deuteron bombardment) by the simple process of heating successive small samples of bombarded LiF in vacuum on a platinum strip; the beryllium diffused into the platinum. When this collection of Be^7 was complete, the strip was heated practically to the melting-point of platinum to bring a fraction of the beryllium to the surface—a process analogous with the 'activation' of a thoriated tungsten thermionic cathode by brief heating at a high temperature. After cooling, the surface was found to emit ions having kinetic energies up to a maximum of 48 volts, and these were interpreted as recoil ions of Li^7; note that they would be expected to leave the surface as ions, not atoms, since the ionization potential of lithium is less than the electron work function of platinum. The detection of the ions was by means of an electron multiplier, to the first cathode of which they were accelerated by a voltage of 3·6 kV., after passing through a pair of grids to which small voltages in either direction could be applied to determine the energies with which the ions left the source.

To meet the objection that some of the K-capture transitions go to an excited state of Li^7 so that the ions might be due to γ-ray recoil, Allen used a γ-ray counter and verified that there was no correlation between the impulses from this counter and those from the multiplier. In any case the maximum energy expected for γ-ray recoil was only about 15 eV.

The experiments of Crane and Halpern, Jacobsen and Kofoed-Hansen, and Allen together add considerable support to the correctness of Pauli's neutrino hypothesis, but it is natural to seek for some further evidence of the neutrino after its emission from the nucleus. It has been suggested that the great flux of neutrinos from the β-active fission products in a nuclear chain-reacting pile might be sufficient to provide observable examples of the very rare interaction of free neutrinos with nuclei, even in the presence of nuclear reactions due to the neutrons and γ-rays. A review of the possibilities was given by Pontecorvo in a lecture that, so far as the writer knows, has not been published, but is contained in a (non-secret) report, *PD* 205, of the Division of Atomic Energy, National Research Council of Canada.

β-transitions from metastable states †

If the lifetime of an excited nucleus is sufficient, a direct transition from the excited state to some state of a neighbouring isobar may be possible; this will, of course, involve the emission of a negatron or a positron, or the capture of an orbital electron. If the nucleus concerned is radioactive in its ground state, obviously such a transition is *energetically* possible since there is available the additional energy of excitation of the metastable state. Whether the direct β-transition from the metastable state will in fact take place with observable intensity depends upon its competition with the isomeric change to the ground state and hence upon the selection rules governing the two competing processes. Since the metastability of the excited state depends essentially on a large difference of spin between it and the ground state, and since β-decay is also improbable if the spin difference between the initial and final states is large, we must not expect the β-transitions from the ground and metastable states to go to the same state of the product nucleus, but otherwise we can make no general prediction.

It should be noted that the β-transitions from the metastable and ground states will, except by accident, have different decay periods: this is to be contrasted with alternative β-transitions from the *same* state of the initial nucleus to different states of the final nucleus, where the initial nucleus decays with a single period (p. 5).

For a discussion with reference to UX_2, see *PRS*, 165, 530. Ti^{51}, Mn^{52} and Lu^{176} are other examples.

Ti^{51}.

This isotope shows two β^--transitions; one has a half-period of 6 minutes and gives 1·6 MeV. β^--rays,‡ the other emits 0·36 MeV. β^--rays, together with a 1 MeV. γ-ray of the product

† See chapter IV.

‡ I.e., the upper energy limit of the continuous spectrum is 1·6 MeV.

nucleus V^{51}, and decays with a 72-day half-life. Both isomers can be produced from Ti^{50} by the (d, p) or (n, γ) reactions.

From Table 2 and Fig. 12 (p. 47), the former is probably an allowed, the latter a first-forbidden transition. The spin of V^{51} is known to be 7/2, so in order that the 1·6 MeV. transition shall be allowed, we must suppose the spin of the metastable state of Ti^{51} to be 5/2, 7/2 or 9/2. The Gamow-Teller selection rules (p. 53) are here assumed. If it were 5/2 or 7/2, the metastable state (in spite of its rather low energy of 0·24 MeV.) would be expected to decay by *isomeric transition* (p. 86) in less than 6 minutes—even if the parity change were unfavourable and the ground state had the lowest possible spin $\frac{1}{2}$; hence the metastable state probably has spin 9/2 and the ground state $\frac{1}{2}$. The 72-day β-transition being first forbidden, it is likely that the excited state of V^{51} has spin 3/2 or 5/2, and either of these values would be consistent with a practically instantaneous 1 MeV. γ-transition to the ground state.

Mn^{52}.

The reaction $Cr(p,n)Mn$ leads to two positron-emitting bodies of periods 21 minutes and 6·5 days (*PR*, 54, 391) with maximum positron energies of 2·2 and 0·77 MeV. respectively (*PR*, 58, 929).† Neither period can be obtained by deuteron bombardment of chromium, but both can be by the (d, α) reaction from iron. They are therefore ‡ presumed to be Mn^{52} isomers. Both positron transitions are accompanied by γ-rays, and the 6·5-day period shows many more γ-rays than positrons. It is therefore to be concluded that this isomer decays to an excited state(s) of Cr^{52} mainly by electron capture and only to a small extent by positron emission. The 21-minute decay proceeds mainly if not entirely by β^+-emission, as would be expected from its higher energy of transition.

Lu^{176}.

This nucleus is sufficiently nearly stable (half-life $\sim 7 \times 10^{10}$ years) to be found in nature, but has a 3·7-hour metastable state that can be made from the natural isomer by neutron or X-ray excitation and decays by β^--emission to Hf^{176}.

† The latter value is probably too high; see p. 59.

‡ The reader may care to fill in the details of the argument, given that the only known stable isotopes are: Cr 50, 52, 53, 54; Fe 54, 56, 57 and 58.

CHAPTER IV

RADIOACTIVE PROCESSES IN WHICH *Z* DOES NOT CHANGE

A nucleus in an excited state can spontaneously change to a state of the same nucleus,† but of lower energy, either by emitting a γ-ray with an energy $h\nu$ equal to the difference between the energies of the two nuclear states, or by giving this energy to an electron in the K, L_I, . . . shell of the same atom, the electron being ejected with kinetic energy $h\nu - E_K$, or $h\nu - E_{L_I}$, etc., where E_K, E_{L_I}, . . . are the binding energies of the K, L_I, . . . electrons.‡

This latter process is called 'internal conversion', the name having arisen by way of analogy with the ordinary photoelectric conversion of the energy of the γ-ray in an atom other than the one from which the γ-ray was emitted. It was originally supposed that the energy of the nuclear transition was first emitted as γ-radiation, the γ-ray then ejecting a photoelectron just as it would from any other atom through which it might pass. Taylor and Mott (*PRS*, 142, 215), however, showed that the process of internal conversion is not a secondary but a primary one, and is due to the direct interaction of the nucleus with its surrounding electrons. This is more than an academic point, for it means that the probability per unit time of the decay of the excited state is not simply the probability of γ-emission, but is the sum of the probabilities of the two competing processes of γ-emission and internal conversion, each of which has to be calculated separately on its own merits. We shall see later that the probability of internal conversion may sometimes be greater than that of γ-emission.

The ejection of a conversion electron creates a vacancy in the inner electronic structure of the atom, and is therefore immedi-

† It may change to a state of *another* nucleus without change of Z by emitting a neutron: this is not classed as a radioactive process, though, in rare instances, it may follow as a consequence of a β^- transition (p. 39).

‡ Internal pair production, i.e. the conversion of the γ-ray energy into the mass- and kinetic energy of a positron-electron pair, is another possibility that has long been recognized, but has not been much studied, and presents puzzling discrepancies between the results of cloud-chamber and counter experiments (see, e.g., *HPA*, 19, 431).

ately followed by the emission of X-rays of the element in question. The observation of these X-rays from a source of small thickness can confirm the presence of internal conversion: if, however, the source is surrounded by any considerable thickness of material (either the same element or another), γ-rays from the source will eject photoelectrons from that material, and the corresponding X-rays will result.

Lifetimes of excited states

The mean lifetime of an excited nucleus before its spontaneous transition to a lower state by γ-radiation or internal conversion depends on many factors. Most transitions take place in a time so short that it can be estimated, if at all, only by indirect arguments; yet some have a half-life measured in minutes, days or months.

It is now fairly certain that these long-lived excited states do not differ essentially from the normal excited states of very short life, and that there must exist many undiscovered examples of intermediate lifetime; but a nucleus in such a metastable state may have quite different radioactive properties from those of the same nucleus in the ground state, and for practical purposes a metastable state may resemble the other states only in possessing the same mass number and nuclear charge: in fact, the existence of long-lived metastable states was first suggested by Weiszäcker to explain the occurrence of such pairs of 'isomeric' nuclei. The term 'isomeric transition' is still generally used to describe the transition (whether by radiation or by internal conversion) of a metastable state to a lower state of the same nucleus. As will be discussed later, metastability is believed to be due to a large difference of angular momentum between the state in question and all lower states of the nucleus.

An important point of nomenclature arises here. Gamma-rays or conversion electrons that follow a radioactive change without measurable time delay are usually listed as 'belonging' to the parent rather than the product nucleus of the disintegration: delayed (isomeric) transitions are, however, always quoted in terms of the nucleus in which they actually occur. This inconsistency will be avoided in the present monograph: the nucleus in which the transition actually occurs will always be specified, the radioactive parent (if any) being also mentioned when it seems desirable to do so. An example may make the point clearer. I^{127} can capture a neutron giving an excited state of I^{128}: this reverts (immediately) to the ground state with the emission of γ-rays *of* I^{128}; the ground state is β-active, transforming with 25 minutes half-life to Xe^{128}, and some of the β-ray transitions go to an excited state of that nucleus, being immediately followed by a γ-ray which is part of the spectrum *of* Xe^{128}.

The complete gamma-ray spectrum

Transitions which follow a radioactive change involve only the few lowest states of the product nucleus; this is because the radioactive change, which normally starts from the ground state of the parent nucleus, cannot for energetic reasons lead to a highly excited state of the product nucleus. In order to observe γ-ray transitions starting from higher states of excitation, it is necessary to obtain the highly excited nucleus as the result of a nuclear reaction.

Owing to the closeness of spacing of the upper levels of excitation, a highly excited nucleus has many states of slightly lower energy to which it can go, and transitions between these upper states will produce many low-energy lines, the probability of any one line being small. A transition direct from a high state to one of the lowest states is more probable than any one of the above low-energy transitions because of the greater energy available, but it is low compared with the sum of their probabilities. Finally, the only lines of high intensity are those resulting from transitions among the lowest states, on which the many alternative upper transitions converge. A purely diagrammatic illustration is given in Fig. 19, which shows hypothetical transitions between one highly excited state and various lower ones.

It should be clear from this discussion that the complete γ-ray spectrum, starting from a highly excited level, will consist of many faint low-energy lines together with the few strong transitions that can, in any case, be observed when the initial excitation is low.

There is the additional experimental difficulty that the upper transitions, if they are to be observed at all, must be observed in the presence of the particles, or radiations, that are producing the nuclear reaction in which the highly excited nucleus is formed: for all these reasons there is practically no experimental information about the upper γ-ray transitions of nuclei. It has been shown, however (see *N*, 139, 323), that the Cd^{113} (n, γ), Cd^{114} reaction yields a considerable number of Cd^{114} gamma-rays.

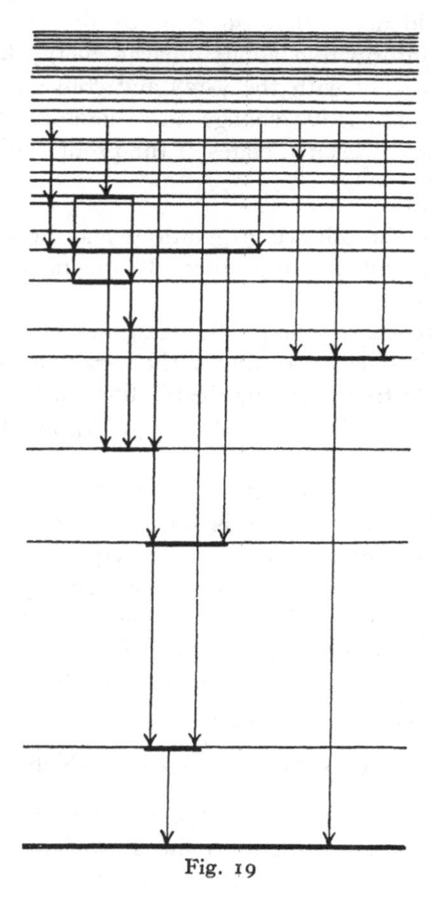

Fig. 19

As Cd^{114} is a stable nucleus, there is in this case no complication due to any subsequent radioactive change.

Information about the lower excited states of nuclei can often be obtained from a detailed study of nuclear reactions: for example, a reaction of the (n, p) type, produced by a mono-energetic neutron source, may give two distinct energy-groups of protons. One group corresponds to the product nucleus being formed in its ground state, and the less energetic group corresponds to this product being left in an excited state: the difference between their energies is equal to the energy of this state. Such processes are, however, beyond the scope of this book, and the experimental evidence which will now be discussed—leading to a good deal of information about nuclear energy levels—will be concerned exclusively with the γ-ray and conversion transitions, which follow a radioactive change and therefore start from one or more of the lower excited states of the product of such change.

Experimental data

Some or all of the following quantities may be measured for a γ-ray transition and for the internal-conversion process with which it competes.

1. The difference of energy between the two nuclear levels concerned. This is equal to $h\nu$ for the γ-ray and to $E + E_s$ for the conversion electron, E_s being the binding energy of the electron in the atom. E_s has different values E_K, E_{LI}, . . . for the K, L_I, ... shells.

2. The half-life associated with the transition, in those instances (isomeric transitions) where it is long enough to be measured. The corresponding decay constant, λ, is equal to the sum of the probabilities per unit time of the various competitive modes of decay of the upper level: it is usually necessary to consider only the partial transition probabilities (decay constants) for γ-radiation, K-shell conversion, and L-shell conversion. If these are denoted by λ_γ, λ_K, λ_L, then $\lambda = \lambda_\gamma + \lambda_K + \lambda_L$.

3. The conversion coefficient, $\dfrac{\lambda_K + \lambda_L}{\lambda_\gamma}$, which is equal to the ratio of the rate of emission of conversion electrons to the rate of emission of γ-rays.† Sometimes the K-shell conversion factor λ_K/λ_γ, and the L-shell conversion factor λ_L/λ_γ, are measured separately. Conversion in the M and other outer shells can usually be neglected.

4. The ratio λ_K/λ_L of conversion in the K- and L-shells.

Of the four quantities listed, the transition energy is fairly readily determined: the half-life is usually immeasurably short, but when measurable at all is measurable with certainty and

† The coefficient is about equally often defined as

$$\frac{\lambda_K + \lambda_L}{\lambda_\gamma + \lambda_K + \lambda_L}, \text{ with partial coefficients } \frac{\lambda_K}{\lambda_\gamma + \lambda_K + \lambda_L}, \frac{\lambda_L}{\lambda_\gamma + \lambda_K + \lambda_L}.$$

accuracy. The conversion coefficient is more difficult to determine: even when the numbers of gamma-rays and conversion electrons are not too different, their comparison rests on the results and interpretation of a considerable number of experiments. If, for example, the excited nucleus is the product of a β-ray change, the conversion ratio is usually estimated by comparing the ratio (conversion electrons)/(β-rays) with the ratio (γ-rays)/(β-rays), and the latter of these ratios can be determined with certainty for an individual γ-ray transition only if a thorough analysis of the γ-ray spectrum has been made. The ratio of conversion in the K- and L-shells, however, can often be measured with reasonable certainty if not with very high accuracy, for it is only necessary to compare the intensities of two lines in the same electron spectrum.

Theoretical considerations

Before discussing the results of actual experiments, we shall find it useful to summarize the theoretical information which can act as a guide to the interpretation of these results.

We first discuss the probability per unit time of γ-ray emission. The quantum theory of radiation utilizes the classical conception of a radiation source as an oscillating electric or magnetic moment, but replaces 'the rate of radiation of energy of frequency ν' by '$h\nu$ times the probability per unit time that a quantum of energy $h\nu$ shall be emitted'. The classical device of analysing a complicated spatial distribution of oscillating charges and currents into spherical harmonics of order 1, 2, 3, . . . L, . . . is also employed, the corresponding electric and magnetic moments being called dipole, quadrupole, octopole, etc., and the same names being also applied by a loose extension to the radiation emitted.

Superimposed on this adaptation of classical electrodynamics is a system of selection rules. They are based upon the following consequences of quantum mechanics.

(i) If the *difference* of angular momentum between the two nuclear states is $L\,\dfrac{h}{2\pi}$, the lowest order multipole that can contribute to the transition is a 2^L-pole.†

† L is the *change* of the total angular momentum of the nucleus, and if atomic notation were to be preserved, it would have to be denoted by ΔJ. The abandonment of the increment symbol Δ is convenient if slightly risky. One thinks of L as a multipole order rather than an angular momentum change. $J = 0 \rightarrow J = 0$ is forbidden.

(ii) A quantum state is said to be odd (−) or even (+) according to whether a reversal of sign of all co-ordinates does or does not reverse the sign of the wave-function describing that state.† If the two states are both odd or both even, then the transition between them involves only electric multipoles of even order and magnetic multipoles of odd order. If the two states are of opposite parity (one being even and one odd), electric multipoles of odd order and magnetic multipoles of even order can alone contribute to the transition.

Thus the lowest allowed multipoles of the two types are as shown in Table 3.

Table 3.

	L even	L odd
Parity of states equal (+, + or −, −)	Electric 2^L-pole Magnetic 2^{L+1}-pole	Electric 2^{L+1}-pole Magnetic 2^L-pole
Parity of states opposite (+, − or −, +)	Electric 2^{L+1}-pole Magnetic 2^L-pole	Electric 2^L-pole Magnetic 2^{L+1}-pole

As will be described below, the radiation probability per unit time decreases rapidly as the multipole order increases, so the lowest allowed multipole is in practice the only one to contribute appreciably to the transition, except that an electric 2^{L+1}-pole can be a serious competitor to a magnetic 2^L-pole. We may therefore finally summarize the selection rules by saying that equal parity and even L, or opposite parity and odd L, lead to electric 2^L-pole radiation. Other combinations give magnetic 2^L and/or electric 2^{L+1}.

Once the magnitudes of the various multipole moments of a given excited nucleus are known, the rate of decay of such nuclei by γ-radiation is easy to calculate; but very different values of multipole moments are obtained with different nuclear models.

The 'liquid drop' model is regarded as the most satisfactory at present available; motions of individual protons are not considered, but the excited nucleus is treated as a continuous electrified sphere traversed by a system of waves of 'thermal' distortion. A statistical estimate can be made of the way the total energy of excitation E is distributed among the various possible

† The question of parity does not arise in one-body problems (e.g. the optical spectra of hydrogen-like atoms) because the parity is then determined by the angular momentum.

frequencies and among the various types of vibration—linear oscillation of the centre of charge, equivalent to an electric dipole; † oscillation between prolate and oblate distortion of the sphere, giving electric quadrupole radiation; and the various higher spherical harmonics. Rotational motions radiate magnetically.

The multipole moments derived from such a treatment take no account of the particular features of individual types of nucleus, and these are likely to be particularly important for the widely-spaced lower excited states that give the prominent lines of observed γ-ray spectra. Formulæ obtained in this way can therefore give only the general dependence of transition probabilities upon energy of excitation, atomic number and so forth.

Following Bethe (*RMP*, 9, 69; § 87), various authors have calculated transition probabilities using the liquid drop model; as an example we may quote the formula

$$\lambda_\gamma = \frac{3Z^2e^2}{4AMR^2c} \cdot \frac{L+1}{1^2 \cdot 3^2 \ldots (2L+1)^2} \cdot \left(\frac{2\pi RE}{hc}\right)^{2L} \quad (12)$$

given by Lowen (*PR*, 59, 835) for an electric 2^L-pole transition of energy E. Here Z is the atomic number, A the mass number of the nucleus and R the nuclear radius; M is the mass of a proton. Lowen's notation has been altered to that of the present monograph, and the last factor is $\left(\frac{2\pi R}{\lambda}\right)^{2L}$ (λ = wavelength).

Since R varies roughly as $A^{\frac{1}{3}}$, the first factor is proportional to $\frac{Z^2}{A^{5/3}}$, which varies very little from one nucleus to another and may be treated as constant. Putting $R = 1{\cdot}5 \times 10^{-13} A^{\frac{1}{3}}$ cm., $Z^2/A^{5/3} = 0{\cdot}8$, we may simplify Lowen's formula to

$$\lambda_\gamma = 1{\cdot}2 \times 10^{20} \frac{L+1}{1^2 \cdot 3^2 \cdot (2L+1)^2} (7{\cdot}8 \times 10^{-3} A^{\frac{1}{3}} E)^{2L} \quad (13),$$

with E in MeV.

Values of λ_γ are shown in Fig. 20 for $A = 27$ (say, Al^{27}) and for $A = 216$ (typical heavy nucleus); if the gamma-ray transition in question is the only mode of decay of that excited state, the half-period is $0{\cdot}693/\lambda_\gamma$ and is indicated by the right-hand scale. The decay period is seen to depend very strongly upon L, which

† The centre of mass of the whole nucleus cannot oscillate, so an electric dipole can only occur if the protons move with respect to the neutrons. All magnetic multipoles also require this, unless the 'drop' has rigidity.

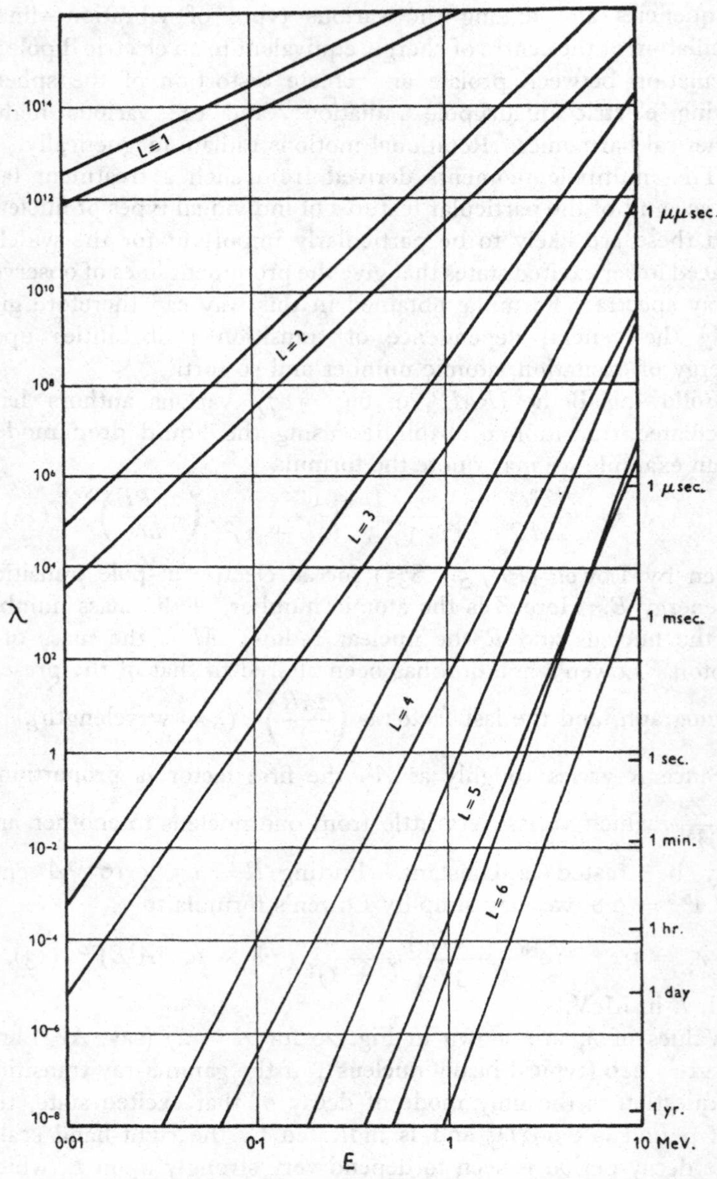

Fig. 20.—Gamma-ray transition probabilities for electric multipoles. The upper line of each pair is for A = 216, the lower for A = 27.

is the number of units of angular momentum by which the initial and final states differ; it is immeasurably short for $L = 1$ or 2, and can be very long for values of L greater than about 3.

For a magnetic multipole of order L, the transition probability may be taken as roughly equal to that for an electric multipole of order $L + 1$.

Since the estimation of multipole moments is the uncertain part of calculations of λ_γ, and since calculations of the K-shell and L-shell conversion probabilities $\frac{\lambda_K}{\lambda_\gamma}, \frac{\lambda_L}{\lambda_\gamma}$ involve the interaction of these same nuclear multipoles with the K and L electrons, it would be expected that the 'conversion ratios' $\frac{\lambda_K}{\lambda_\gamma}$ and $\frac{\lambda_L}{\lambda_\gamma}$ could be calculated more simply and reliably than the individual transition probabilities λ_γ, λ_K and λ_L.

Unfortunately the results are complicated and differ considerably according to the types of approximation involved (non-relativistic for light nuclei and low energies; fully relativistic for heavy nuclei and high energies). The conversion ratios vary rapidly with Z, and it seems impossible to present both clearly and concisely the quantitative results obtained by various authors for different ranges of atomic number and transition energy. Generally, both the K and L conversion ratios are predicted to increase with increasing Z and L and with decreasing E. Below a transition energy that equals the ionization energy of the K-shell, no K conversion can, of course, occur; but Hebb and Uhlenbeck (*P*, 5, 605), who first calculated conversion ratios for high values of angular momentum change L, point out that the L-shell conversion ratio, normally somewhat smaller than that for the K-shell, will continue to increase as the available energy decreases from E_K to E_L.

A few typical values of the K-shell conversion coefficient as calculated by Hebb and Uhlenbeck for $Z = 41$ are given in Table 4. It will be seen that for the low energies and high values of L that give long life against γ-emission the conversion coefficient is a good deal larger than one; that is to say, long-lived ('metastable') excited nuclei will be expected to decay mainly by internal conversion and not by γ-radiation, at least if the energy of transition is low. For energies of the order of 0·5 MeV. and upwards, γ-emission should become predominant even for $L = 5$.

Table 4.—Conversion ratios (i.e. ratios of K-shell conversion-electron emission to gamma-ray emission) predicted by Hebb and Uhlenbeck for $Z = 41$.

Table 4.

Gamma-ray Energy (MeV.)	$L = 2$	$L = 3$	$L = 4$	$L = 5$
0·05	9·1	90	850	7800
0·10	0·8	5·6	37·6	251

For heavier nuclei and for greater energies, see, e.g., *PR*, 58, 486, where conversion of magnetic multipole radiation is also considered.

The ratio of K to L conversion is, as has been mentioned already, much easier to determine experimentally than are the individual L and K conversion ratios, and, though usually less sensitive to the multipole order, is an equally good, perhaps better, basis for deducing the multipole order of an observed transition. In general, it is necessary to use both methods in order both to determine L and to settle the question of whether an electric or a magnetic multipole is acting. Several examples are considered by Helmholz (*PR*, 60, 415), and others will be referred to in the present chapter.

Typical experimental methods and results

We shall describe experiments on a cascade of γ-ray transitions in Xe^{130}, each transition being, so far as is known, of immeasurably short delay, and therefore associated with a small change in angular momentum of the Xe^{130} nucleus. We have already mentioned a somewhat similar cascade of γ-rays; those of Cr^{52}. This nucleus can be produced by the β^--decay of V^{52} and also by the β^+-decay of Mn^{52}; the latter change leaves the nucleus in a higher state of excitation than does the former, and thus gives more lines of the Cr^{52} spectrum. (See p. 71 and Fig. 16.)

Xe^{130}: γ-rays and conversion electrons (Roberts, Elliott, Downing, Peacock and Deutsch, *PR*, 64, 268)

Sources of radioactive I^{130} were prepared by bombarding Te (in the form of cobalt telluride) with 11·5 MeV. deuterons: the target was dissolved in nitric acid, with the addition of a small amount of inactive iodine to act as carrier; the iodine was separated by distillation and ultimately used in the form of NaI or AgI. Other active isotopes of iodine have considerably shorter or considerably longer half-periods; their effects were small and easily corrected for.

The β-ray spectrum of a thin sample, taken with a magnetic lens spectrometer, showed four faint but definite conversion lines at values of $B\rho$ (about 2·5, 3·0, 3·4 and 3·7 kilogauss-cm.) corresponding to electron energies of 0·382, 0·504, 0·631 and 0·713 MeV. The electrons come from the K-shell of Xe^{130}, for which the binding energy is 0·034 MeV.; the transition energies come out to be 0·416, 0·538, 0·665 and 0·747 MeV.

When the source was enclosed in a brass capsule thick enough to absorb the I^{130} β-rays (max. energy 1·03 MeV.), a spectrum of Compton recoil electrons was obtained showing only two broad groups (broken line of Fig. 21): a gold foil placed on top of the capsule gave, superimposed on the Compton spectrum, four very clear peaks and three smaller peaks at values of $B\rho$ corresponding respectively to Au K-shell photoelectrons from γ-rays of energies 0·417, 0·535, 0·670 and 0·740 MeV., and to L-shell photoelectrons from three of these γ-rays. The fourth L-shell line could not be observed because it happens to fall on one of the K-shell lines. There is thus no doubt of the interpretation of the internal conversion spectrum.

The relative numbers of quanta of these four energies can be estimated from the relative heights of the photoelectron peaks above the Compton electron background. This is possible because experimental † and theoretical ‡ values, in good agreement with one another, are available for the variation with γ-ray energy of the photoelectric yield for lead: the variation must be very similar for gold. It was found that three of the gamma-rays were equally numerous: that is, the ratio $\dfrac{\text{observed number of photoelectrons}}{\text{known photoelectric yield per }\gamma\text{-ray}}$ was the same for all three. The 0·417 MeV. γ-rays, however, were only about $(30 \pm 10)\%$ as numerous.

Now the β-ray spectrum of I^{130} is complex, somewhat more

† Gray, *PCPS*, 27, 103. ‡ Hulme *et al.* *PRS*, 149, 131.

than half the disintegrations having a maximum energy of 1·03 MeV., and the remainder having a maximum energy of 0·61 MeV. The difference between these energies is 0·42 MeV. and therefore equal, within experimental error, to the energy of that Xe^{130} γ-ray which is less intense than the others. Clearly it was tempting to conclude that the disintegration scheme is as shown

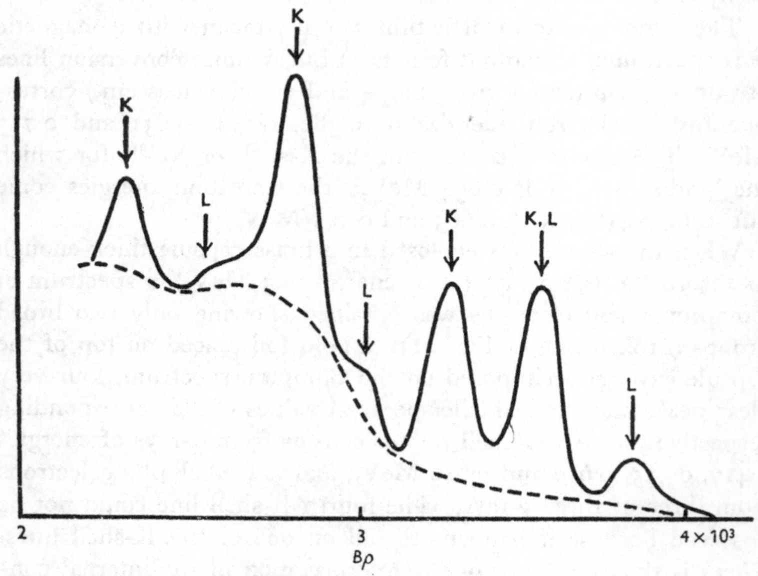

Fig. 21.—Photoelectron spectrum from gold irradiated with Xe^{130} γ-rays (Roberts *et. al.*). The photoelectron peaks are seen above a background of Compton electrons (broken line).

in Fig. 22 (*a*), where the order of the last three transitions is unknown and is indicated quite arbitrarily. A possible alternative would have been the scheme of Fig. 22 (*b*), though the energy balance could then only just be brought within the combined limits of error of the various energy measurements involved.

To decide between these two schemes, various types of coincidence experiment were performed. The number of $\beta\gamma$ coincidences per recorded β-ray was first investigated as a function of the thickness of a celluloid absorber placed between the source and the *beta*-ray counter. It fell from about 5×10^{-3} with no absorber to about $4{\cdot}2 \times 10^{-3}$ with an absorber sufficiently thick to absorb all the β-rays of the low-energy group. Since the net efficiency of the γ-ray counter varied from $0{\cdot}90 \times 10^{-3}$ to $1{\cdot}06 \times 10^{-3}$ for the various γ-ray energies in question, it was clear that each β-ray is accompanied by several γ-rays. There was the obvious temptation to divide $4{\cdot}2 \times 10^{-3}$ by an average γ-ray counting efficiency of, say, $1{\cdot}0 \times 10^{-3}$ and deduce that the high-energy β-ray is accompanied by about four γ-rays; the low-energy β-ray would then be accompanied by five or more.

However, experiments on $\gamma\gamma$ coincidences gave somewhat lower

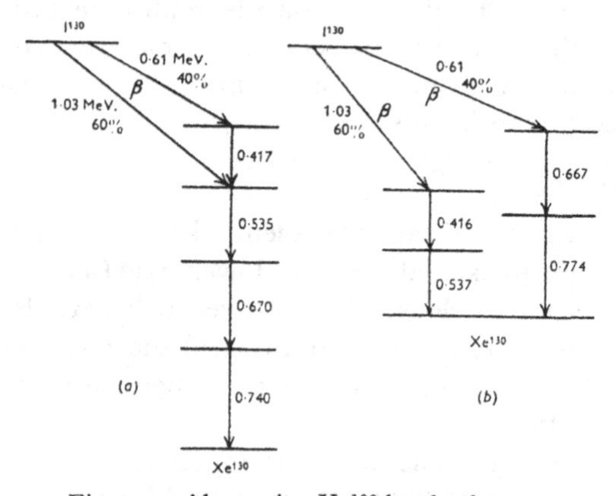

Fig. 22.—Alternative Xe^{130} level schemes

values. In these, the number of $\gamma\gamma$ coincidences was divided by the number of single γ-ray counts; this ratio was compared with the formulæ given on p. 25 for the case where *two* γ-rays are emitted in cascade, and was also compared with similar (naturally more complicated) formulæ for cascades of three, four, . . . γ-rays. The results showed that each γ-ray is on the average accompanied by more than two and less than three other gamma-rays.

These coincidence studies clearly eliminated the scheme of Fig. 22 (*b*), so the scheme of Fig. 22 (*a*) may be taken as certain apart from the order of the three lowest transitions.

The internal conversion coefficients for the four transitions were measured by comparing the number of electrons recorded in each conversion line with the number of beta-rays leading to the transition in question: the coefficients found were 12×10^{-3}, $6{\cdot}9 \times 10^{-3}$, $3{\cdot}8 \times 10^{-3}$, $3{\cdot}1 \times 10^{-3}$ in order of increasing transition energy. These values agree remarkably well with an extrapolation to $Z = 54$ of the theoretical results for electric quadrupole transitions of Hulme *et al.*: the non-relativistic calculations of Hebb and Nelson give somewhat poorer agreement, particularly for the more energetic transitions. Even if the assignment to electric quadrupole transitions proves incorrect, the fact that the four conversion coefficients all lie on a smooth curve when plotted against the respective transition energies may be taken to indicate that the radiations are of the same type. The K/L conversion ratio could not be used to decide the multipole order since (in agreement with theoretical expectation for radiations of low multipole order) the L conversion was found to be too small to be detected in these experiments.

Isomeric transitions

About forty instances are at present † known of delayed nuclear transitions from an excited state to a lower state (usually the ground state) of the same nucleus; the observed half-periods range from less than a minute to several months, and there is every reason to suppose that many shorter and some longer periods will in due course be identified.

The first clearly established example of such an 'isomeric transition' was that of Br^{80}. It was found by Fermi *et al.* (*PRS*, 149, 522) that the irradiation of bromine by slow neutrons yields three apparently distinct β^--ray bodies of half-periods 18 minutes, 4·4 hours and 34 hours: the carrier of each of these activities is chemically identifiable as bromine. This was a surprising observation, since there are only two stable isotopes of bromine, Br^{79} and Br^{81}, and in this region of the periodic table the normal result of the entry of a slow neutron into a nucleus is the capture of the neutron with the formation of the next higher isotope of the element in question. It might have been argued, of course, that one of the three periods resulted from the formation of Br^{78} by an $(n, 2n)$ process: but the removal of a neutron from a nucleus requires the provision of energy and cannot be achieved by the impact of a *slow* neutron on Br^{79}. When energy was supplied to bromine nuclei by the irradiation of bromine by 17 MeV. γ-rays

† This was written in the Spring of 1947; by August 1948 the total has risen to over fifty.

(from the nuclear reaction $Li^7 + p \rightarrow Be^8 + \gamma$), Bothe and Gentner (*ZP*, 106, 235) again found three bromine activities, of half-periods 6·4 minutes, 18 minutes and 4·5 hours: the first yields positrons, the last two negatrons, just as do the 18-minute and 4·4-hour activities discovered by Fermi.

Bothe and Gentner concluded that the 18-minute and 4·5-hour bromines were both Br^{80}, being produced in their experiments by the reaction Br^{81} (γ, n) Br^{80}, and in Fermi's experiments by the reaction Br^{79} (n, γ) Br^{80}.

If indeed any confirmation of this conclusion were necessary, it was supplied by Fleischmann, who found that the 18-minute and 4·5-hour periods were both produced by the selective (resonance) capture by bromine of neutrons of the same energy—viz., about 50 eV.

The relation between these two forms of Br^{80} can be inferred from a number of experiments, of which the most striking are certain applications of the Szilard-Chalmers process, by Segrè, Halford and Seaborg and by De Vault and Libby.

1. The β^--activity from bromine after a short irradiation by neutrons decays with time according to a somewhat complex law, finally settling down to an exponential of 4·5-hour half-life, but the energy of the β-rays does not vary during the decay, its upper limit of about 2 MeV. in particular remaining unchanged.

2. If the radioactive bromine has remained for some time in the form of (e.g.) C_2H_5Br, or of a bromate ion in aqueous solution, then free bromine atoms or ions are evolved, which may be separated from the parent substance by appropriate physical or chemical means (shaking the C_2H_5Br with water or precipitation of AgBr from the bromate solution). The β^--activity of the separated bromine decays with the single period of 18 minutes, while the unseparated bromine has lost its β-ray activity; it recovers at a rate characteristic of a half-period of 18 minutes, and a further extraction of the 18-minute activity can then be made. Clearly, the 18-minute bromine is the true β-ray emitter and is the product of a 4·5-hour isomeric change during which the Br^{80} atom becomes separated from the organic molecule or the bromine ion of which it forms part. The fact that, even immediately after a short irradiation of bromine with neutrons, the 18-minute activity is present, and in considerable strength, is

explained by the direct production of the 18-minute body, as well as the 4·5-hour body, in the initial nuclear reaction. As this original source of 18-minute nuclei decays, it is partly replaced by the growth of fresh 18-minute nuclei from the 4·5-hour isomeric transition.

The process may thus be symbolized as below:

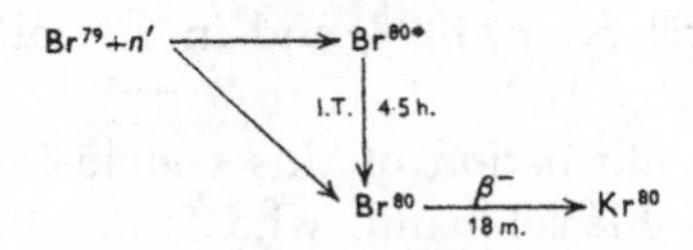

Here the metastable state is indicated by an asterisk.

Confirmation of this conclusion may be obtained in various ways: for example by detection of the primary and secondary phenomena accompanying the isomeric transition. The primary phenomena are of course the emission of γ-rays and conversion electrons: they have been studied by several investigators and perhaps most extensively by Berthelot (*AP*, 19, 117 and 19, 219). The isomeric transition to the ground state of Br^{80} appears to proceed by way of an intermediate excited state, so that either two conversion electrons (and no γ-ray quantum), or one quantum and one electron, are emitted in the de-excitation of each metastable nucleus. Berthelot made a careful analysis of the absorption in aluminium foils of the Br^{80} β-rays and the Br^{80*} conversion electrons, as emitted from the mixed source decaying in radioactive equilibrium: a similar investigation of the β-ray spectrum of Br^{80} alone, as separated by the Szilard-Chalmers method, showed that no conversion electrons of low energy ($<$ 0·6 MeV.) are emitted by Br^{80}, so any found from the mixed source must be ascribed to Br^{80*}. Berthelot found discontinuities of slope of the 'mixed' absorption curve which corresponded to the ranges of the groups of electrons of energies 47·2 keV., 35·5 keV. and 23·6 keV. which had previously been identified by Valley and McCreary (*PR*, 55, 666; 56, 863). The first represents the *L*-conversion of a 49-keV. quantum, the second the *K*-conversion of the same quantum plus the *L*-conversion of a 37-keV. quantum, the third the *K*-conversion of the 37-keV. quantum. By comparing the number of electrons in each group with the number of Br^{80}

β-rays,† making use of the fact ‡ that the only γ-ray actually emitted is of 37-keV., the 49-keV. transition being therefore '100% internally converted', Berthelot deduced the following partial conversion coefficients (defined according to the footnote of p. 76).

Table 5.

	49 keV. transition	37 keV. transition
K	0·88	0·34
L	0·12	0·05

By comparison of these conversion coefficients with the theoretical values given by the 'liquid drop' nuclear model, Berthelot deduces that the 49-keV. transition is a magnetic octopole, and the 37-keV. transition a magnetic dipole. The theoretical half-life for the former comes out to be about 2 hours, for the latter about 10^{-8} sec.: the 49-keV. transition is therefore identified as the long-delayed transition with the observed half-life of 4·5 hours, and the 37-keV. transition must always follow it in cascade, since no transition of energy $49 + 37 = 86$ keV. is observed.

The transition scheme is therefore as shown below.

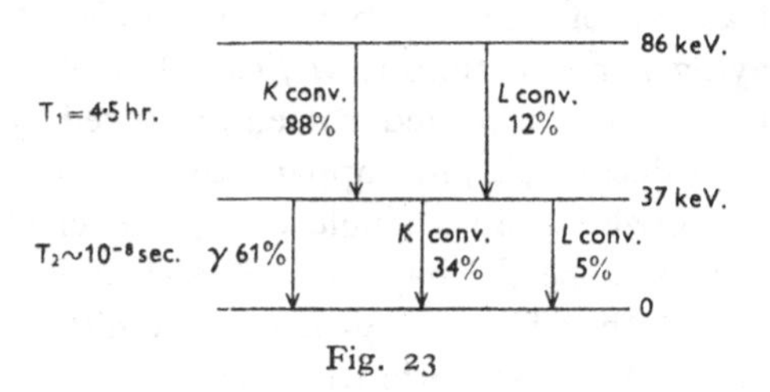

Fig. 23

If the mean life of the 37-keV. excited state were longer than a few microseconds, a coincidence-counting system of high resolution would show only chance coincidences between conversion electrons: actually, a system of resolving time 2·3 microseconds recorded genuine coincidences, proving that $T_2 < 2{\cdot}3 \times 10^{-6}$ sec.

† Allowance must be made for the fact that the Br^{80*} source has decayed slightly between the instant when a conversion electron is observed and the instant when the β-ray from the same atom is observed.

‡ Grindberg and Roussinow (*PR*, 58, 8).

The expected secondary effects of the internal conversion process are firstly, bromine X-rays and, secondly, electrons (' Auger electrons ') due to the internal conversion of these X-rays in the L-, M-, . . . shells of the bromine atoms concerned. The bromine K-line has been duly identified, by the absorption-edge method,† by Roussinow and Yousephowitch, while the Auger electrons have been observed by Siday and by Berthelot.

It is presumably a matter of chance that the half-period of the isomeric transition is longer than that of the subsequent β-decay, and the fact that it happens to be so in nearly all the well-established examples of metastable states of radioactive nuclei can be explained on experimental grounds. If the isomeric transition has the shorter period, it is usually difficult to detect its γ-rays in the presence of those arising from the decay of the ground state, which are usually both more energetic and more numerous. One of the few examples where a short-lived isomeric transition has been established in the presence of a longer-lived β^--decay from the ground state is provided by Y^{91}, where the isomeric transition has a half-period of 50 minutes as compared with the 57-day β^--decay: it is here possible to observe the isomeric transition (*a*) because it has a *very* much shorter period than the β^--decay, so that even if the initial number of metastable nuclei is a small fraction of the number of ground-state nuclei, the number decaying per unit time is *at first* relatively much greater, (*b*) because the nucleus concerned is itself produced by the β^--decay of Sr^{91}, and a β-decay can (and apparently does in this instance) go with high probability to a single excited level of the product nucleus, and (*c*) because Y^{91} happens to go entirely to the ground state of Zr^{91} and therefore no γ-rays are emitted to mask the rather soft γ-rays of the isomeric transition.

' Instantaneous ' γ-rays from higher excited states of Y^{91} may be eliminated by chemical separation of the active yttrium from its parent strontium. The chain of events could be represented by Fig. 24, which is, however, hypothetical so far as the details of the Sr^{91} decay is concerned.

The examples so far considered are of isomeric transitions in which the ground state is β-active: in some instances, however, the ground state is stable—as for example with In^{115} and with

† p. 27.

In^{113}. Although, as for β-active nuclei, the metastable and ground states may here be formed side by side as the result of a nuclear reaction or a radioactive change, it is also possible in such cases to produce the metastable state by *nuclear excitation*, irradiating

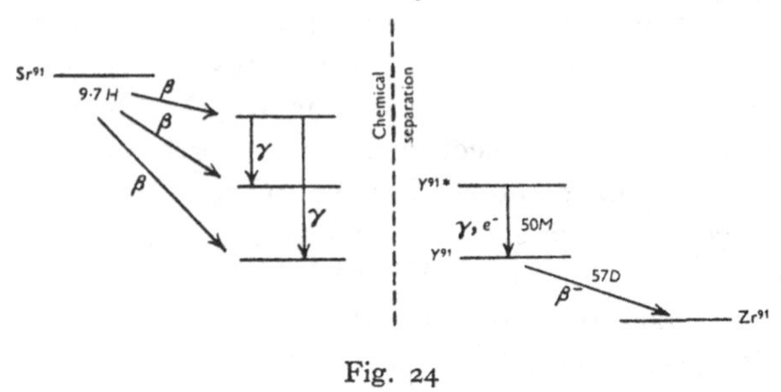

Fig. 24

the naturally-occurring ground-state nuclei with sufficiently energetic X-rays, or neutrons. By inelastic collisions, the nuclei are raised in *allowed* transitions to states higher than the metastable one, from which higher states they, or some of them, drop down to the metastable state. The process is illustrated schematically in Fig. 25, from which it will be clear that if we ascribe the

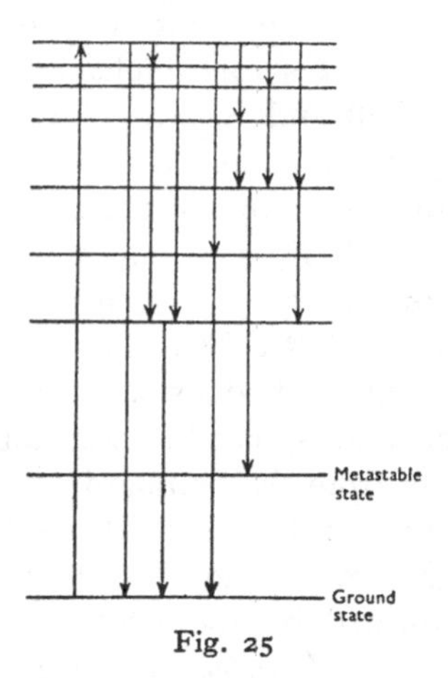

Fig. 25

metastability to a large difference of angular momentum, we must expect to have to bring the nucleus to a comparatively high level of excitation in order that it may get to the metastable state by several successive small changes of spin: even so, a return to the ground state without passing through the metastable state will be much more probable on grounds of angular momentum change, and we must expect the yield of metastable nuclei to be low.

The nuclear excitation of indium by X-rays was first reported by Pontecorvo and Lazard (*CR*, 208, 99) and subsequently studied by Collins *et al.* (*PR*, 55, 507). The threshold is at 1·35 MeV. and the principal product is 4·5-hour In^{115*}, which gives both γ-rays and conversion electrons. Dunworth and Pontecorvo (*PCPS*, 43, 123) have shown that the metastable state of the less abundant isotope In^{113} can also be produced by 2-MeV. X-rays. So, also, can the β^--active Lu^{176*} (*PCPS*, 43, 429).

Fast electrons can also cause nuclear excitation, and various metastable species have been formed through this process (*PR*, 57, 1088; 67, 92).

Bombardment of stable nuclei by fast neutrons can similarly cause nuclear excitation: when the nucleus concerned has a metastable state, sufficiently high nuclear excitation will result in a fraction of the nuclei being 'caught' in that state, just as with X-ray excitation. Cross-sections for nuclear excitation by fast protons (i.e. for inelastic scattering of fast protons by nuclei) are likewise enough (10^{-29} cm.^{2}) for the production of readily-observable numbers of metastable nuclei in suitable cases. For example, Goldhaber, Hill and Szilard (*PR*, 55, 47) and Barnes and Aradine (*PR*, 55, 51) obtained the 4·1-hour metastable In^{115} by neutron and proton excitation respectively.

Dore and Pontecorvo (*CR*, 208, 99) report that a metastable Cd isotope can be produced by fast neutron bombardment of Cd, while the formation of Rh^{103*} from the only stable rhodium isotope, Rh^{103}, has been studied by a number of workers. An activity of about one hour's half-life was found by Pool, Cork and Thornton to result from the bombardment of Rh by neutrons from the Li (d, n) process: it was much weaker than that of the well-known 44-sec. β^--emitter Rh^{104} or of its 4·2-minute isomer, both of which are formed by the Rh^{103} (n, γ) Rh^{104} reaction. Flammersfeld (*Naturwissenschaften*, 32, 36) and Hole (*AMAF*, 32A, 3) have studied what is apparently the same activity: according to Hole the radiation from Rh^{103*} consists of soft γ-rays of about 19 ± 3 keV., decaying with a 53-minute half-period.

Statistical survey of isomeric transitions

Two general rules regarding the existence of metastable nuclear states were pointed out by Mattauch (*ZP*, 117, 250), and each has a simple explanation in terms of the belief that the half-life against γ-emission or internal-conversion transitions *increases* with increasing change of angular momentum and *decreases* with increasing energy of transition.

The first rule is that no metastable states are observed in nuclei having even numbers of both neutrons and protons. The reason is that the ground state of such a nucleus, with the protons and neutrons arranged in pairs of opposite spin, has zero spin and a very low energy compared with that of even the first excited state, which is probably one MeV. or more removed: the second and third excited states may also be quite widely spaced, and there is no chance of finding an excited state of the high spin and moderate energy which are both required for metastability of appreciable lifetime. For somewhat similar reasons (large level spacing, small spin values) the nuclei of atomic number below 20 do not show isomerism.

Mattauch's second rule is that all nuclei with a ground-state spin of 9/2 have a metastable excited state: this is plausible since 9/2 is an exceptionally high value, and the first excited state is very likely to have the most common value for a nucleus of odd A: viz., $\frac{1}{2}$. One would, however, not expect this rule to be *invariably* true, since the first excited state might well have a spin of 3/2 or 5/2, giving spin difference of 3 or 2 units which, as we shall see later, are too small to account for half-lives of seconds or more. Bi^{209} seems to be an exception to the rule. It may be added that no metastability has yet been detected for nuclei of which the ground-state spin is known to be 5/2 or 3/2: several instances occur with nuclei whose ground-state spin is $\frac{1}{2}$, and this is natural because a first-excited state of spin 9/2 must be expected to occur occasionally.

According to the calculations mentioned on p. 79, the transition probability, while depending very strongly on the angular momentum and energy changes involved, should depend only moderately upon the atomic number, principally by way of the nuclear radius. It is therefore permissible to compare experimental values of transition probabilities and transition energies with these predicted by theory for a chosen moderate value of Z (say 40), and to see whether rough agreement between theory and experiment can be obtained for reasonable values of L, the angular momentum change. Such a comparison was made by Berthelot (*AP*, 19, 117, Fig. 5), who has given curves showing the logarithm of the half-period for γ-transitions *alone* plotted against the logarithm of the transition energy, as calculated from a formula essentially the same as that of Bethe, taking $Z = 35$ and $A = 80$, and giving L the successive values 3, 4, 5, 6. Of the seven isomeric transitions for which experimental data were available (i.e. for which the effect of internal conversion on the half-life was either absent or could be allowed for from experimental results) six lay fairly close to the theoretical curve for $L = 5$ and one lay close to the curve for $L = 4$.

These groupings of isomeric transitions, which have an obvious analogy with the Sargent groupings for isobaric (β^+, β^-) transitions, have been further examined by Wiedenbeck (*PR*, 69, 567), who has included the *theoretical* correction for the increased transition probability due to internal conversion (p. 82): he has plotted the *total* transition probability and the transition energy logarithmically against one another for various versions of the theory and has included the experimental points for about forty isomeric transitions.

Fig. 26 is a replot of Wiedenbeck's collection of data, with additional material, using Lowen's recalculation of Bethe's 'liquid drop' formula, and making allowance for internal conversion. The broken lines indicate the theoretical predictions for various values of L for 'medium' nuclei. On the whole, the evidence for the existence of two distinct groups is fairly strong, and their identification with angular momentum changes of 4 and 5 units is at least plausible. Two points (for Nb^{93} and element 61) lying to the right of $L = 5$ might possibly correspond to $L = 6$, but neither is very well established experimentally; nor are most of those few which fall about midway between the main two groups, though Sr^{85} is believed to be beyond suspicion. The subject deserves a more critical examination of the theoretical and the experimental data, and a less crude way of comparing them, than has here been given.

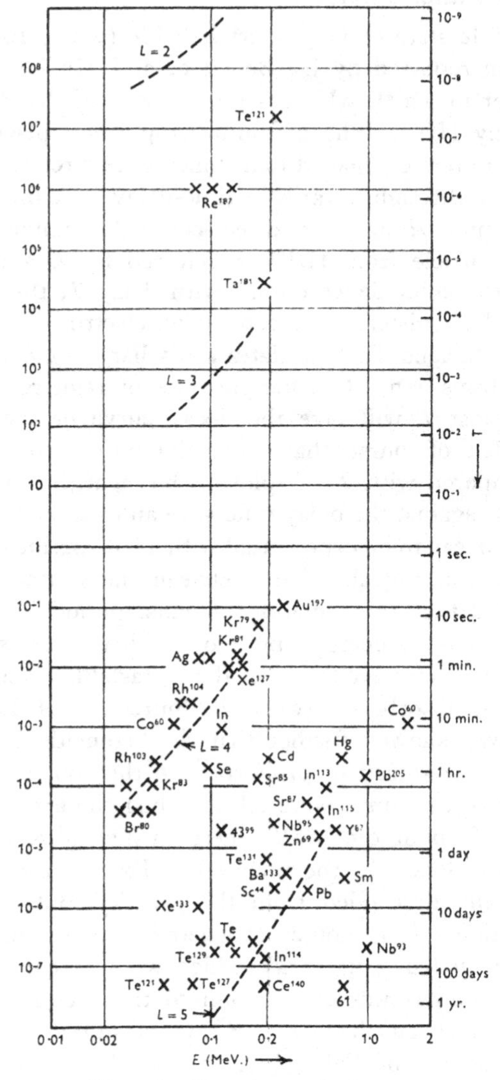

Fig. 26.—Energies and decay constants of isomeric transitions

Short-lived metastable states

A metastable state of very short half-life ($2{\cdot}2 \times 10^{-5}$ sec.) has recently been reported by De Benedetti and McGowan (*PR*, 70, 569) to occur in Ta^{181}, which is the product of the β^--transition of the 55-day Hf^{181}. The hafnium sample was placed between two β-ray counters connected to a coincidence circuit, one directly and the other through a variable time-delay. Calling the delay-coupled counter A and the direct-coupled counter B, we see that if a β-particle from Hf^{181} is detected by A and passes an impulse to the coincidence circuit with delay T, then a 'coincidence' will be registered if a conversion electron (or γ-ray) from the product nucleus Ta^{181} initiates a discharge in B at a time T after the β-transition. Plotting the rate of registration of coincidences against T will give the decay curve of the metastable state, provided, of course, that the resolving time τ of each counter is short compared with T. A plot of the logarithm of the 'coincidence' rate against the delay time gave an excellent straight line, showing the decay to be exponential with a half-period of 22 microseconds. By placing absorbers between the source and B, the authors found that the radiation of $22\mu s.$ period is mostly conversion electrons of energy about $0{\cdot}11$ MeV., corresponding to $0{\cdot}18$ MeV. energy of excitation of the metastable level.

By inserting absorbers between the source and A, De Benedetti and McGowan showed further that the 'coincidences' between A and B were reduced in proportion to the single β-ray counts in A; after these β-rays were all absorbed there remained some γ-radiation of about $0{\cdot}5$ MeV. energy which gave counts in A but no 'coincidences': the metastable Ta^{181} nuclei were therefore genetically descended from the β-decay, but not from the γ-ray transition. It is not clear whether the γ-radiation occurs in the Hf^{181} nucleus (Fig. 27*a*) (which would mean that the 55-day period is an isomeric transition to the β-emitting level) or is associated with an alternative β-ray transition to a non-metastable excited level of Ta^{181} (Fig. 27*b*) or whether the metastable level of Ta^{181} decays to an intermediate excited state which then proceeds by 'prompt' γ-radiation to the ground state (Fig. 27*c*). These alternatives could be tested by ordinary coincidence experiments with $T = 0$ and an absorber (sufficient to absorb all β-rays) between the source and one of the two counters. If the first supposition is true, no coincidences would be expected: the second and third would give $(\gamma\beta)$ and (γe) coincidences respectively. Later work (*PPS*, 61, 300) confirms Fig. 27 (*c*).

It will be seen from Fig. 26 that the relation between the half-life and transition energy agrees fairly well with the theoretical prediction for a change of three units in the nuclear spin; that is, for electric octopole radiation, which requires the parities of the two levels concerned to be opposite. The alternative would be magnetic quadrupole, with no change of parity, the spin change

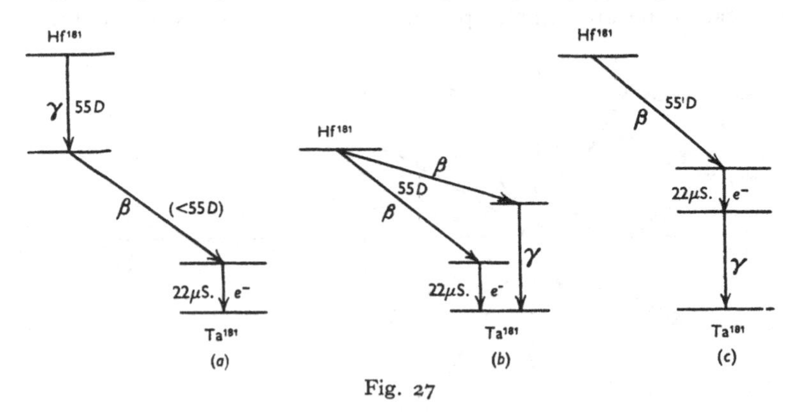

Fig. 27

being then two units of $h/2\pi$. The ground state of Ta^{181} is known to have spin 7/2.

Two still shorter-lived metastable states (Re^{187}, $\sim \frac{1}{2} \times 10^{-6}$ sec., *PR*, 71, 380 and Te^{121}, $\sim 5 \times 10^{-8}$ sec., *PR*, 70, 780) have since been reported; they seem to fall midway between the lines for $L = 2$ and $L = 3$. The short-lived state of Te^{121} is itself the product of a higher metastable state of the same nucleus, having a half-life of over 100 days. An even more recent example, Ge^{72}, ($\frac{1}{2} \times 10^{-6}$ sec.; *PR*, 72, 1219) is unusual because it is an 'even-even' nucleus.

Angular correlation of successive γ-rays.

It had long been suspected (see, e.g., *RSI*, 11, 167) that there might be some correlation in angle between the directions of emission of cascade γ-rays from the same nucleus. How this comes about may be seen (ignoring photon spin) from the example of a nucleus, originally having zero angular momentum, that undergoes two successive electric dipole transitions and finally has again zero angular momentum.

Let the first γ-ray leave the nucleus (O) in the direction $O\gamma_1$ (Fig. 28). The nucleus is left with an angular momentum $h/2\pi$, the axis of which must be perpendicular to $O\gamma_1$; take a typical direction of this axis perpendicular to the plane of the paper. The second γ-ray removes this angular momentum of the nucleus, and must therefore go out in the plane of the paper; all directions in that plane are equally probable, as indicated by the equally-

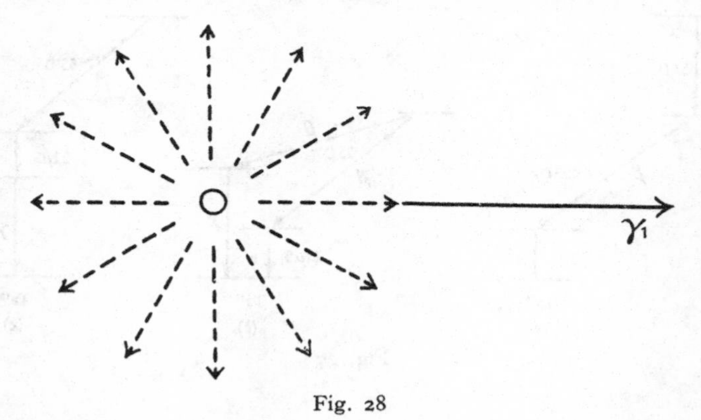

Fig. 28

spaced arrows in the figure. Since all directions of nuclear spin perpendicular to $O\gamma$ are equally probable, the complete spatial distribution of probability is obtained by rotating the diagram about $O\gamma$. As may be seen by viewing the diagram obliquely, from the bottom of the page, the most probable directions of emission of the second γ-ray are $O\gamma$ and the opposite direction γO, with a minimum probability at right angles. The actual probability distribution is of the form $(1 + \cos^2\theta)$, where θ is the angle between the two γ-rays.

If the second transition is long enough delayed for the nucleus to change its axis of spin (by precession in an external magnetic field, or in that of its own electronic structure), the correlation diminishes. Most other sequences of nuclear angular momentum, and other multipole orders, give less marked correlations; in particular, there is no correlation at all if the intermediate nuclear state has zero spin, for the nucleus has then no means of 'remembering' its previous history. Until recently, attempts to observe such correlations experimentally had failed; but they have now been observed for Co^{60} and Sc^{46} (*PR*, 72, 870) and will doubtless be found in other nuclei. This phenomenon gives promise of becoming a most useful tool in the investigation of the spins of nuclear excited states, and of the multipole orders of transitions between them. It has been pointed out that similar correlations are likely to exist between the directions of successive conversion electrons (duly found; *N*, 163, 168) and between the polarisation directions of successive γ-ray photons (*PR*, 73, 518), as well as between the directions of a β-ray and an immediately-following γ-ray (*PR*, 73, 649).

INDEX OF SUBJECTS

Printed in the United States
By Bookmasters